AMERICA IN THE 21ST CENTURY: POLITICAL AND ECONOMIC ISSUES

STEM (SCIENCE, TECHNOLOGY, ENGINEERING, AND MATHEMATICS) WORKFORCE TRENDS AND POLICY CONSIDERATIONS

America in the 21st Century: Political and Economic Issues

Additional books in this series can be found on Nova's website under the Series tab.

Additional e-books in this series can be found on Nova's website under the e-books tab.

Economic Issues, Problems and Perspectives

Additional books in this series can be found on Nova's website under the Series tab.

Additional e-books in this series can be found on Nova's website under the e-books tab.

AMERICA IN THE 21ST CENTURY: POLITICAL AND ECONOMIC ISSUES

STEM (SCIENCE, TECHNOLOGY, ENGINEERING, AND MATHEMATICS) WORKFORCE TRENDS AND POLICY CONSIDERATIONS

CLARK MARTIN GARCIA
AND
SCOTT COOK
EDITORS

NOVINKA
New York

For permission to use material from this book please contact us:
Telephone 631-231-7269; Fax 631-231-8175
Web Site: http://www.novapublishers.com

LIBRARY OF CONGRESS CATALOGING-IN-PUBLICATION DATA

ISBN: 978-1-62257-818-4

Published by Nova Science Publishers, Inc. † New York

CONTENTS

PREFACE

Science, technology, engineering and mathematics (STEM) workers drive our nation's innovation and competitiveness by generating new ideas, new companies and new industries. However, U.S. businesses frequently voice concerns over the supply and availability of STEM workers. Over the past 10 years, growth in STEM jobs was three times as fast as growth in non-STEM jobs. STEM workers are also less likely to experience joblessness than their non-STEM counterparts. Science, technology, engineering and mathematics workers play a role in the sustained growth and stability of the U.S. economy, and are a critical component to helping the U.S. win the future. This book explores workforce trends and policy considerations for STEM students and workers.

Chapter 1 - Although the United States remains the leading host country for international students in science, technology, engineering, or mathematics (STEM) fields, the global competition for talent has intensified. A record number of STEM graduates—both U.S. residents and foreign nationals—are entering the U.S. labor market, and there is a renewed focus on creating additional immigration pathways for foreign professional workers in STEM fields. Current law sets an annual worldwide level of 140,000 employment-based admissions, which includes the spouses and children in addition to the principal (i.e., qualifying) aliens. "STEM visa" is shorthand for an expedited immigration avenue that enables foreign nationals with graduate degrees in STEM fields to adjust to legal permanent resident (LPR) status without waiting in the queue of numerically limited LPR visas. The fundamental policy question is should the United States create additional pathways for STEM graduates to remain in the United States permanently? The number of full-time graduate students in science, engineering, and health fields who were foreign students (largely on F-1 nonimmigrant visas) grew from 91,150 in

1990 to 148,923 in 2009, with most of the increase occurring after 1999. Despite the rise in foreign student enrollment, the percentage of STEM graduate students with temporary visas in 2009 (32.7%) was comparable to 1990 (31.1%). Graduate enrollments in engineering fields have exhibited the most growth of the STEM fields in recent years. After completing their studies, foreign students on F-1 visas are permitted to participate in employment known as Optional Practical Training (OPT), which is temporary employment that is directly related to an F-1 student's major area of study. Generally, a foreign student may work up to 12 months in OPT status. In 2008, the Department of Homeland Security (HHS) expanded the OPT work period to 29 months for F-1 students in STEM fields. Many F-1 visa holders (especially those who are engaged in OPT) often change their immigration status to become professional specialty workers (H-1B). Most H-1B beneficiaries are typically admitted to work in STEM occupations. In FY2010, the most recent year for which detailed data on H-1B beneficiaries (i.e., workers renewing their visas as well as newly arriving workers) are available, almost 91,000 H-1B workers were employed in computer-related occupations, and they made up 47% of all H-1B beneficiaries that year. The H-1B visa and the OPT often provide the link for foreign students to become employment-based LPRs. In total, foreign nationals reporting STEM occupations made up 44% of all of the 676,642 LPRs who were employment-based principal immigrants during the decade of FY2000- FY2009. Of all of the LPRs reporting STEM occupations (297,668) over this decade, 52% entered as professional and skilled workers. STEM graduates seeking LPR status are likely to wait in line to obtain LPR status. Those immigrating as professional and skilled workers face wait times of many years, but those who meet the criteria of the extraordinary ability or advanced degrees preference categories have a much shorter wait. There may be renewed interest in establishing STEM visas in the 112th Congress, and several bills (H.R. 399, H.R. 2161, H.R. 3146, S. 1965, and S. 1986) have been introduced. The House Committee on the Judiciary held two hearings on these issues in 2011. These issues also arose during a 2011 Senate Committee on the Judiciary hearing on the economic rationale for immigration reform. No legislation on STEM visas, however, has moved through committees thus far.

Chapter 2 - Science, technology, engineering and mathematics (STEM) workers drive our nation's innovation and competitiveness by generating new ideas, new companies and new industries. However, U.S. businesses frequently voice concerns over the supply and availability of STEM workers.

Over the past 10 years, growth in STEM jobs was three times as fast as growth in non-STEM jobs.

STEM workers are also less likely to experience joblessness than their non-STEM counterparts. Science, technology, engineering and mathematics workers play a key role in the sustained growth and stability of the U.S. economy, and are a critical component to helping the U.S. win the future.

- In 2010, there were 7.6 million STEM work- ers in the United States, representing about 1 in 18 workers.
- STEM occupations are projected to grow by 17.0 percent from 2008 to 2018, compared to 9.8 percent growth for non-STEM occupations.
- STEM workers command higher wages, earning 26 percent more than their non- STEM counterparts.
- More than two-thirds of STEM workers have at least a college degree, compared to less than one-third of non-STEM workers.
- STEM degree holders enjoy higher earnings, regardless of whether they work in STEM or non-STEM occupations.

Chapter 3 - Our science, technology, engineering and math (STEM) workforce is crucial to America's innovative capacity and global competitiveness. Yet women are vastly underrepresented in STEM jobs and among STEM degree holders despite making up nearly half of the U.S. workforce and half of the college-educated workforce. That leaves an untapped opportunity to expand STEM employment in the United States, even as there is wide agreement that the nation must do more to improve its competitiveness.

- Although women fill close to half of all jobs in the U.S. economy, they hold less than 25 percent of STEM jobs. This has been the case throughout the past decade, even as college-educated women have increased their share of the overall workforce.
- Women with STEM jobs earned 33 percent more than comparable women in non-STEM jobs – considerably higher than the STEM premium for men. As a result, the gender wage gap is smaller in STEM jobs than in non-STEM jobs.
- Women hold a disproportionately low share of STEM undergraduate degrees, particularly in engineering.
- Women with a STEM degree are less likely than their male counterparts to work in a STEM occupation; they are more likely to work in education or healthcare.

- There are many possible factors contributing to the discrepancy of women and men in STEM jobs, including: a lack of female role models, gender stereotyping, and less family-friendly flexibility in the STEM fields. Regardless of the causes, the findings of this report provide evidence of a need to encourage and support women in STEM.

Chapter 4 - Science, technology, engineering, and math (STEM) workers are essential to American innovation and competitiveness in an increasingly dynamic and global marketplace. In this third report, we examine demographic disparities in STEM education and find that educational attainment may affect equality of opportunity in these critical, high-quality jobs of the future.

This report follows an analysis of labor market outcomes and gender disparities among STEM workers. We find that regardless of race and Hispanic origin, higher college graduation rates are associated with higher shares of workers with STEM jobs. But non-Hispanic Whites and Asians are much more likely than other minority groups to have a bachelor's degree. By increasing the numbers of STEM workers among currently underrepresented groups through education we can help ensure America's future as a global leader in technology and innovation.

Other key findings of this report include:

- Non-Hispanic Whites comprise the largest group of STEM workers, accounting for about seven out of ten STEM workers, which aligns closely with their share of the overall workforce.
- Non-Hispanic Asians are most likely (42 percent) to graduate college with a STEM degree, while the propensities of other groups are all fairly similar (17-22percent).
- Half of all non-Hispanic Asian workers with STEM degrees have STEM jobs, compared to 30 percent of Hispanics and non-Hispanic Black and American Indian and Alaska Native workers.
- One in five STEM workers is foreign-born, of which 63 percent come from Asia.
- STEM workers in all demographic groups, including the foreign-born, earn more than their non-STEM counterparts. Hispanics and non-Hispanic Blacks receive a significantly larger STEM premium than do non-Hispanic Whites.

Chapter 5 - From 1989 through 2009 the number of minority U.S. citizens and permanent residents enrolled in graduate science and engineering (S&E) programs more than doubled, growing from approximately 37,700 in 1989 to 92,700 in 2009. Increases in Hispanic, black, and Asian/Pacific Islander S&E

graduate students were similar over this period (approximately 17,800, 18,200, and 17,200, respectively); however, these gains almost tripled the number of Hispanic graduate students (approximately 190% growth) and more than doubled the number of blacks (approximately 155% growth) and Asians/Pacific Islanders (approximately 110% growth). Enrollment among American Indians/Alaska Natives also nearly tripled, increasing from approximately 900 in 1989 to approximately 2,600 in 2009 (approximately 195% growth). Minority enrollment among U.S. citizens and permanent residents enrolled in graduate S&E programs grew from approximately 13% in 1989 to approximately 24% in 2009 (figure 1). Due to extra variability of the methodological changes in the 2007 Survey of Graduate Students and Postdoctorates in Science and Engineering (GSS), all growth rate calculations comparing pre- and post-2007 counts are rounded to the nearest 5% and counts are rounded to the nearest 100; see "Data Limitations and Availability" for more information. Despite these substantial gains, black and Hispanic U.S. citizens and permanent residents remain underrepresented within the S&E graduate student population when compared with the adult U.S. citizen population. In 2009, 7.8% of the U.S. citizens and permanent residents pursuing graduate S&E degrees were black and 7.1% were Hispanic. In 2009, 13.8% of U.S. citizens 21 to 45 years of age were black and 11.9% were Hispanic. In contrast, the percentage of American Indians/Alaska Natives in the 2009 S&E graduate student population was very similar to that of the adult U.S. citizen population (0.7% vs. 0.8%), and the percentage of Asians/Pacific Islanders pursuing S&E graduate degrees was more than twice that of the U.S. citizen population aged 21 to 45 years (8.6% vs. 3.8%). These and other findings in this Info-Brief are from the fall 2009 GSS, cosponsored by the National Science Foundation (NSF) and the National Institutes of Health (NIH). The GSS is an annual survey of all academic institutions in the U.S. that grant research-based master's degrees or doctorates in science, engineering, or selected health (SEH) fields. The GSS collects data on the number and characteristics of graduate students, postdoctoral appointees (postdocs), and other doctorate-holding nonfaculty researchers in SEH fields.

Chapter 6 – This is Testimony of Darla Whitaker, Senior Vice President, Worldwide Human Resources, Texas Instruments. Hearing on "STEM The Tide: Should America Try To Prevent An Exodus of Foreign Graduates of U.S. Universities with Advanced Science Degrees?"

Chapter 7 - Statement of Vivek Wadhwa, Director of Research, Center for Entrepreneurship and Research Commercialization, Duke University. Hearing on "STEM The Tide: Should America Try to Prevent an Exodus of Foreign Graduates of U.S. Universities with Advanced Science Degrees?"

Chapter 8 - Statement of Lindsay Lowell, Director of Policy Studies, Institute for the Study of International Migration. Hearing on "STEM the Tide: Should America Try to Prevent an Exodus of Foreign Graduates of U.S. Universities with Advanced Science Degrees?"

In: STEM Workforce Trends ... ISBN: 978-1-62257-818-4
Editors: C. Martin Garcia and S. Cook

Chapter 1

IMMIGRATION OF FOREIGN NATIONALS WITH SCIENCE, TECHNOLOGY, ENGINEERING, AND MATHEMATICS (STEM) DEGREES*

Ruth Ellen Wasem

SUMMARY

Although the United States remains the leading host country for international students in science, technology, engineering, or mathematics (STEM) fields, the global competition for talent has intensified. A record number of STEM graduates—both U.S. residents and foreign nationals—are entering the U.S. labor market, and there is a renewed focus on creating additional immigration pathways for foreign professional workers in STEM fields. Current law sets an annual worldwide level of 140,000 employment-based admissions, which includes the spouses and children in addition to the principal (i.e., qualifying) aliens. "STEM visa" is shorthand for an expedited immigration avenue that enables foreign nationals with graduate degrees in STEM fields to adjust to legal permanent resident (LPR) status without waiting in the queue of numerically limited LPR visas. The fundamental policy question is

* This is an edited, reformatted and augmented version of a Congressional Research Service publication, CRS Report for Congress R42530, from www.crs.gov, prepared for Members and Committees of Congress, dated May 11, 2012.

should the United States create additional pathways for STEM graduates to remain in the United States permanently?

The number of full-time graduate students in science, engineering, and health fields who were foreign students (largely on F-1 nonimmigrant visas) grew from 91,150 in 1990 to 148,923 in 2009, with most of the increase occurring after 1999. Despite the rise in foreign student enrollment, the percentage of STEM graduate students with temporary visas in 2009 (32.7%) was comparable to 1990 (31.1%). Graduate enrollments in engineering fields have exhibited the most growth of the STEM fields in recent years.

After completing their studies, foreign students on F-1 visas are permitted to participate in employment known as Optional Practical Training (OPT), which is temporary employment that is directly related to an F-1 student's major area of study. Generally, a foreign student may work up to 12 months in OPT status. In 2008, the Department of Homeland Security (HHS) expanded the OPT work period to 29 months for F-1 students in STEM fields.

Many F-1 visa holders (especially those who are engaged in OPT) often change their immigration status to become professional specialty workers (H-1B). Most H-1B beneficiaries are typically admitted to work in STEM occupations. In FY2010, the most recent year for which detailed data on H-1B beneficiaries (i.e., workers renewing their visas as well as newly arriving workers) are available, almost 91,000 H-1B workers were employed in computer-related occupations, and they made up 47% of all H-1B beneficiaries that year.

The H-1B visa and the OPT often provide the link for foreign students to become employment-based LPRs. In total, foreign nationals reporting STEM occupations made up 44% of all of the 676,642 LPRs who were employment-based principal immigrants during the decade of FY2000- FY2009. Of all of the LPRs reporting STEM occupations (297,668) over this decade, 52% entered as professional and skilled workers. STEM graduates seeking LPR status are likely to wait in line to obtain LPR status. Those immigrating as professional and skilled workers face wait times of many years, but those who meet the criteria of the extraordinary ability or advanced degrees preference categories have a much shorter wait.

There may be renewed interest in establishing STEM visas in the 112th Congress, and several bills (H.R. 399, H.R. 2161, H.R. 3146, S. 1965, and S. 1986) have been introduced. The House Committee on the Judiciary held two hearings on these issues in 2011. These issues also arose during a 2011 Senate Committee on the Judiciary hearing on the economic rationale for immigration reform. No legislation on STEM visas, however, has moved through committees thus far.

INTRODUCTION

Congress is renewing its interest in facilitating the immigration of foreign professional workers in science, technology, engineering, or mathematics (STEM) fields.

The STEM workforce is seen by many as a catalyst of U.S. global economic competitiveness and is likewise considered a key element of the legislative options aimed at stimulating economic growth.[1] "STEM visa" is a shorthand for an expedited immigration avenue that enables foreign nationals with graduate degrees in STEM fields to adjust their immigration status to legal permanent residence (LPR) without waiting in the queue of numerically-limited LPR visas.[2]

The fundamental policy question is: should the United States create additional pathways for STEM graduates to remain in the United States permanently? Or, are current avenues adequate?

The answer to the question lies at the nexus of education policy, labor force needs, and immigration priorities. More precisely, the key elements are: the source countries of international STEM students; the hiring choices of U.S. employers; and, the statutory limits and priorities of U.S. immigration law. This report opens by presenting a statistical portrait of foreign nationals studying STEM fields in U.S. institutions.

An analysis of the current avenues foreign nationals with STEM degrees use to work in the United States temporarily and permanently follows. Discussions of the legislative history, current legislation, and major issues of debate conclude the report.

BACKGROUND

The Immigration and Nationality Act (INA) currently allocates 140,000 visas annually for economic immigrants, of which over 120,000 are allocated to three employment-based preference categories.

These employment-based LPR categories are (1) persons of extraordinary ability in the arts, sciences, education, business, or athletics; outstanding professors and researchers; and certain multinational executives and managers; (2) members of the professions holding advanced degrees or persons of exceptional ability; and (3) skilled workers with at least two years training, professionals with baccalaureate degrees, and unskilled workers in occupations in which U.S. workers are in short supply.[3]

Although the United States remains the leading host country for international students in STEM fields, the global competition for talent, most notably with the European Union and Asian nations, has intensified in recent years. Concerns that these educational and career opportunities foster a "brain drain" of professionals from developing nations are infrequently expressed in the current debate. Instead, some researchers warn of a "reverse brain drain" because the United States does not allocate what they consider to be a sufficient number of visas for high-skilled immigrants to fuel the economy. According to the U.S. Department of Commerce, "growth in STEM jobs was three times as fast as growth in non-STEM jobs" over the past ten years.[4]

Other researchers maintain that a record number of STEM graduates—both U.S. residents and foreign nationals—are entering the U.S. labor market. They express concern that foreign nationals would displace U.S. residents in the STEM fields if additional visas were allocated. Some analysts observe that the only high-skilled occupations that experienced negative wage growth in recent years were technology-related occupations (e.g., computer programmers and engineers)—occupations in which highly-educated foreign nationals cluster. Almost two-thirds of the 9.3 million people in the U.S. labor market who had STEM degrees in 2010 were employed in non-STEM occupations.[5]

Caveat

There is no generally accepted definition of what specific academic disciplines "STEM" encompasses. The National Science Foundation (NSF) studies the fields broadly and includes biological, agricultural, and environmental life sciences; computer and information sciences; mathematics and statistics; the physical sciences; psychology; the social sciences; engineering; and health fields.[6]

For the purposes of tracking foreign students who study in the United States, the Department of Homeland Security's Immigration and Customs Enforcement (ICE) defines STEM more narrowly and does not include economics, sociology, and political science; however, ICE recently expanded its list, adding fields such as animal science, educational statistics, environmental science, nutritional science, and various specializations within psychology.[7] The National Center for Education Statistics often uses the ICE definition.[8]

TEMPORARY FOREIGN STUDENTS

International students pursuing STEM degrees in the United States are part of a long tradition of foreign students studying at U.S. institutions of higher education. Since the Immigration Act of 1924, the United States has expressly permitted foreign students to study in U.S. institutions. While their presence is generally viewed as a positive one, Congress has strengthened the reporting requirements of educational institutions with foreign students.[9] Foreign students are among the various categories of foreign nationals who are known as nonimmigrants.[10]

The most common visa for foreign students is the F-1 visa. It is tailored for international students pursuing a full-time academic education.[11] Spouses and children may accompany the F-1 visa holders on F-2 visas, but are not permitted to work. Students on F visas are generally barred from off-campus employment; however, they are permitted to work in practical training that relates to their degree program, such as paid research and teaching assistantships.[12] The Department of Homeland Security (DHS) Student and Exchange Visitor Information System (SEVIS) database reports 512,884 F-1 students were in active status (i.e., enrolled full-time or otherwise engaged in approved activities) in FY2010.[13]

Foreign Nationals Earning STEM Degrees

The number of full-time graduate students in science, engineering, and health fields who were foreign students grew from 91,150 in 1990 to 148,923 in 2009. Based upon the more broadly-defined NSF data, *Figure 1* shows that most of the increase occurred after 1999. Despite the rise in foreign student enrollment, the percentage of STEM graduate students with temporary visas in 2009 (32.7%) was comparable to 1990 (31.1%). Graduate enrollments in engineering fields have exhibited the most growth of STEM fields in recent years.[14] A snapshot of degrees granted to foreign students by broad categories indicates that engineering is leading field for both MA and Ph.D. recipients in 2010 (Figure 2). Computer science fields follow for MA degrees; while mathematics and physical sciences place second for Ph.D. degrees. Although based upon data from ICE, the STEM fields presented in Figure 2 use the broader NSF definition of STEM, which includes economics, sociology, psychology and political science as well as agricultural and life sciences, computer sciences, mathematics and the physical sciences.

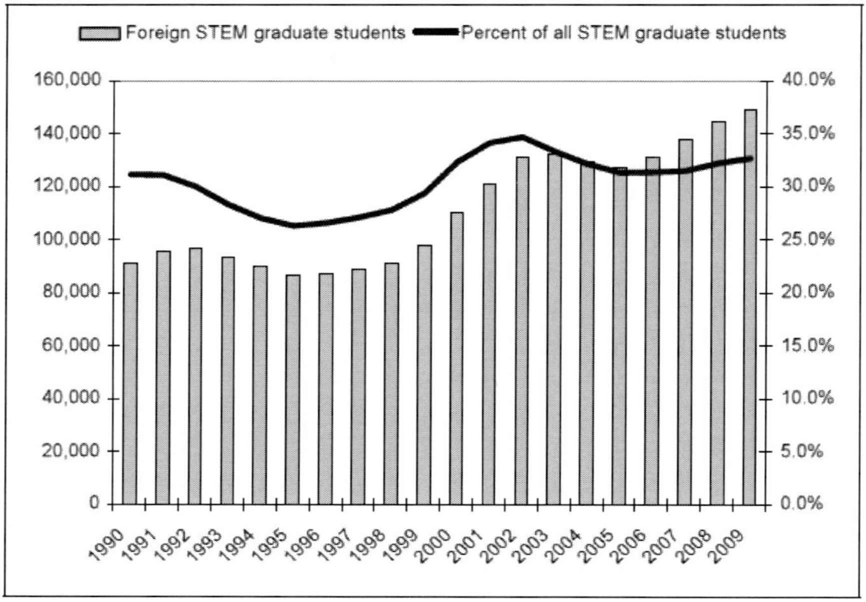

Source: National Science Foundation/National Center for Science and Engineering Statistics, NSF-NIH Survey of Graduate Students and Postdoctorates in Science and Engineering.

Figure 1. Full-Time Graduate Students with Temporary Visas in Science, Engineering, and Health Fields, 1990-2010.

India was the top sending country for STEM graduates earning masters degrees and represented 56% of all STEM graduates earning masters degrees in 2009 (*Table 1*).

Indian graduate students studying engineering and computer science led in masters degrees. Students from China placed second at 15% and were more evenly distributed across STEM fields, with the exception of engineering. The remaining top 8 sending countries accounted for a total of 13% of all foreign students earning masters degrees in STEM fields in 2009, less than that of China.

In terms of doctoral degrees (*Table 2),* China sent the most STEM graduates in 2009, making up 35%. Chinese Ph.D. recipients were almost half of all foreign nationals in mathematics and physical sciences and roughly a third of all foreign nationals in the other STEM fields (except psychology and the social sciences). Doctoral degree recipients from India represented 16% of STEM doctorates in 2009.[15]

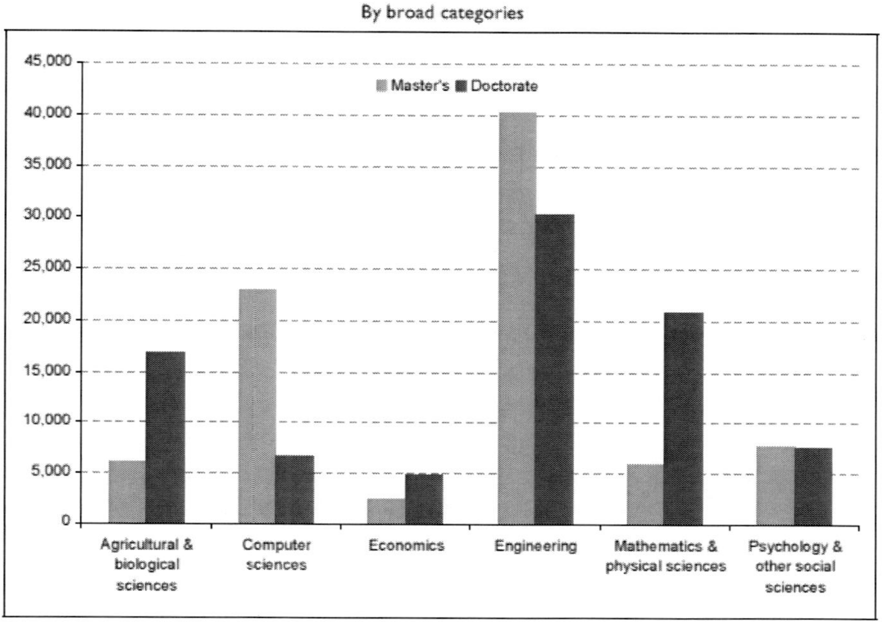

Source: Department of Homeland Security, U.S. Immigration and Customs Enforcement, Student and Exchange Visitor Information System database, special tabulations for the National Science Foundation, 2010.

Note: Although based upon data from ICE, these STEM fields use the broader NSF definition of STEM.

Figure 2. Foreign Nationals Receiving Advanced Degrees in STEM Fields, 2009.

TEMPORARY FOREIGN WORKERS

The Immigration and Nationality Act of 1952 authorized visas for foreign nationals who would perform needed services because of their high educational attainment, technical training, specialized experience, or exceptional ability.

Today, there are several temporary visa categories that enable employment-based temporary admissions for highly skilled foreign workers. They perform work that ranges from professional and skilled labor to those who work in jobs requiring extraordinary ability in the sciences, arts, education, business, or athletics.

Optional Practical Training (OPT)

After completing their studies, F-1 foreign students are permitted to participate in employment known as Optional Practical Training (OPT), which is temporary employment that is directly related to an F-1 student's major area of study. Generally, an F-1 foreign student may work up to 12 months in OPT status.

In 2008, DHS expanded the OPT work period to 29 months for F-1 students in STEM fields.

To qualify for the 17-month extension, F-1 students must have received STEM degrees included on the STEM Designated Degree Program List, be employed by employers enrolled in E-Verify,[16] and have received an initial grant of post-completion OPT related to such a degree (i.e., already approved for 12 months in OPT).[17] According to U.S. Citizenship and Immigration Service (USCIS), the number of F-1 visa holders who are engaged in OPT was 92,465 in FY2010.

Table 1. Master's Degrees in STEM Fields Awarded to Foreign Nationals, 2009 Top 10 sending countries by broad categories

	Agricultural Biological Sciences	*Computer Science*	*Economics*	*Engineering*	*Mathema-tics Physical Sciences*	*Psycholo-gy Social Sciences*	*Percent of Total*
Total Master's	6,050	22,880	2,480	40,300	5,810	7,600	
India	2,750	16,270	140	26,290	1,210	500	56%
China	1,030	2,240	1,040	5,100	2,290	1,220	15%
Taiwan	240	520	100	1,240	200	340	3%
South Korea	140	440	140	910	280	670	3%
Nepal	160	460	50	430	110	100	2%
Saudi Arabia	70	320	40	380	90	110	1%
Canada	150	70	30	210	70	470	1%
Japan	90	70	40	130	70	520	1%
Turkey	30	130	80	380	60	190	1%
Thailand	40	250	40	330	40	90	1%

Source: Department of Homeland Security, U.S. Immigration and Customs Enforcement, Student and Exchange Visitor Information System database, special tabulations for the National Science Foundation, 2010.

Note: Although based upon data from ICE, these STEM fields use the broader NSF definition of STEM.

Table 2. Doctorate Degrees in STEM Fields Awarded to Foreign Nationals, 2009 Top 10 sending countries by broad categories

	Agricultural Biological Sciences	*Computer Science*	*Economics*	*Engineering*	*Mathematics Physical Sciences*	*Psychology Social Sciences*	*Percent of Total*
Total							
Doctorates	16,830	6,650	4,800	30,380	20,900	7,490	
China	5,550	2,410	1,300	10,570	8,680	980	35%
India	3,600	1,250	370	5,690	2,800	530	16%
South Korea	1,080	520	530	3,100	1,240	1,070	9%
Taiwan	930	220	150	1,480	700	400	4%
Turkey	250	210	280	980	440	460	3%
Iran	90	210	60	1,830	270	50	3%
Canada	460	90	80	430	380	690	2%
Japan	240	30	160	160	210	340	1%
Germany	210	70	100	150	250	210	1%
Nepal	200	40	60	160	410	50	1%

Source: Department of Homeland Security, U.S. Immigration and Customs Enforcement, Student and Exchange Visitor Information System database, special tabulations for the National Science Foundation, 2010.

Note: Although based upon data from ICE, these STEM fields use the broader NSF definition of STEM.

They comprise 18% of all active F-1 foreign students in FY2010. STEM graduates performing OPT during their first year after graduation are included among the 82,206 F-1 students shown in *Figure 3* as "Post-Completion." In addition, there were 10,022 F-1 students who obtained the 17-month extension for OPT in STEM fields.

Temporary Professional Specialty Worker: H-1B Visas

Many F-1 visa holders (especially those who are engaged in OPT) often change their immigration status to become professional specialty workers (H-1B). Other H-1B visa holders, however, are adjusting from another immigration status or coming directly from abroad. The law sets numerical restrictions on annual admissions of the H-1B (65,000). Most H-1B workers, however, enter on visas that are exempt from the ceiling because they are returning workers or they work for universities and nonprofit research facilities that are exempt from the cap. The H-1B visas are closely associated with STEM fields, but are not limited to them.

Prospective H-1B nonimmigrants must demonstrate to USCIS that they have the requisite education and work experience for the posted positions. After the Department of Labor (DOL) approves the labor attestation, USCIS processes the petition for the H-1B nonimmigrant (assuming other immigration requirements are satisfied) for periods up to three years. An alien can stay a maximum of six years on an H-1B visa.

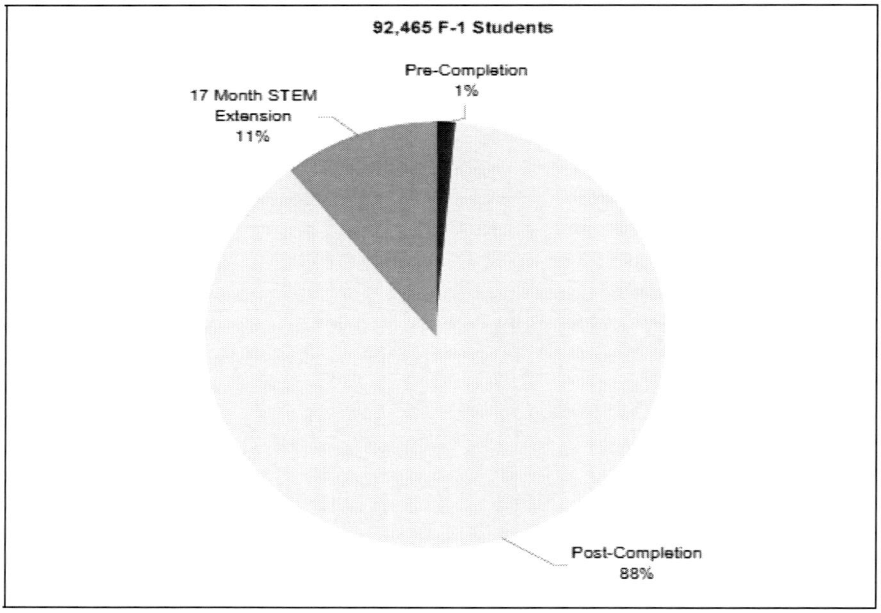

Source: CRS presentation of data from U.S. Citizenship and Immigration Services.
Notes: Post-Completion category includes those STEM graduates in the first year of OPT.

Figure 3. F-1 Foreign Students Performing Optional Practical Training in FY2010.

The H-1B labor attestation, a three-page application form, is the first step for an employer wishing to bring in an H-1B professional foreign worker. The attestation is a statement of intent rather than a documentation of actions taken.[18] In the labor attestation for an H-1B worker, the employer must attest that the firm will pay the nonimmigrant the greater of the actual wages paid to other employees in the same job or the prevailing wages for that occupation; the firm will provide working conditions for the nonimmigrant that do not cause the working conditions of the other employees to be adversely affected; and that there is no applicable strike or lockout. The firm must provide a copy of the labor attestation to representatives of the bargaining unit or—if there is no bargaining representative—must post the labor attestation in conspicuous locations at the work site.[19]

In FY2011, there were 218,500 H-1B professional specialty worker petitions approved, down from a high of 288,000 in FY2004. Although the law sets a numerical limit of 65,000 H-1B workers each year, only initial grants are counted under the cap.

As noted above, there are several categorical exceptions to the cap, most notably all H-1B nonimmigrants who work for universities and nonprofit research facilities. A provision in P.L. 108-447 also exempts up to 20,000 aliens holding a master's or higher degree from the numerical limit on H-1B visas.

As *Figure 4* displays, over the past decade more H-1B workers were approved outside of the numerical cap of 65,000 than under the cap. Not all H-1B workers with approved petitions actually use the visa.

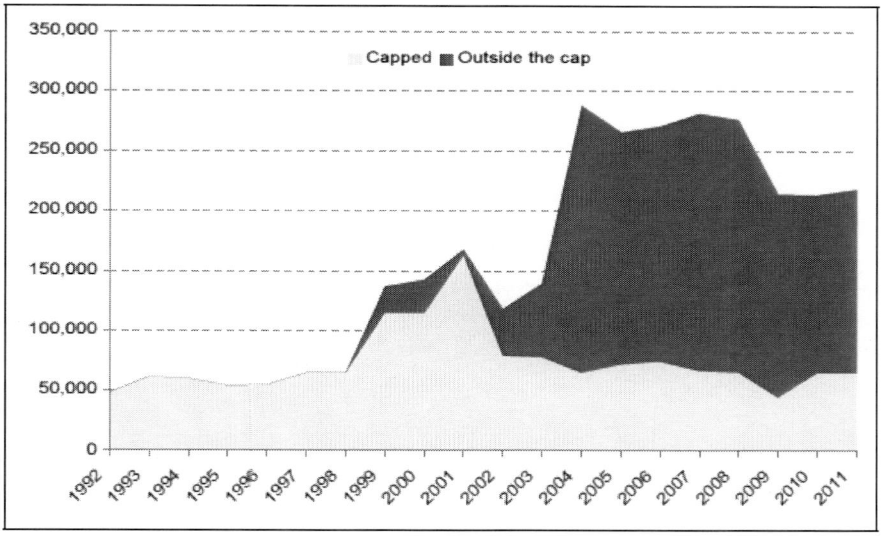

Source: CRS presentation of data from the DHS Office of Immigration Statistics (OIS) and its predecessor in the Immigration and Naturalization Service.

Notes: Congress increased the H-1B cap to 115,000 for FY1999-FY2000 and to 195,000 for FY2001-FY2003.

Figure 4. Total H-1B Petitions Approved, FY1992-FY2011.

Over the years, a noteworthy portion of H-1B beneficiaries have worked in STEM occupations. In FY2010, the most recent year for which detailed data on H-1B beneficiaries (i.e., workers renewing their visas as well newly arriving workers) are available, almost 91,000 H-1B workers were employed in computer-related occupations, and they made up 47% of all H-1B beneficiaries that year, as *Figure 5* indicates. Architectural and engineering occupations as well as occupations in education were tied at a distant second with 10% each. Administrative occupations followed with 9%, and health and medicine occupations were 8% of the 192,990 H-1B beneficiaries. The total number of H-1B beneficiaries reported for FY2010 (192,990) and shown in *Figure 5* was less than the number of approved H-1B petitions approved that year as depicted in *Figure 4*.[20]

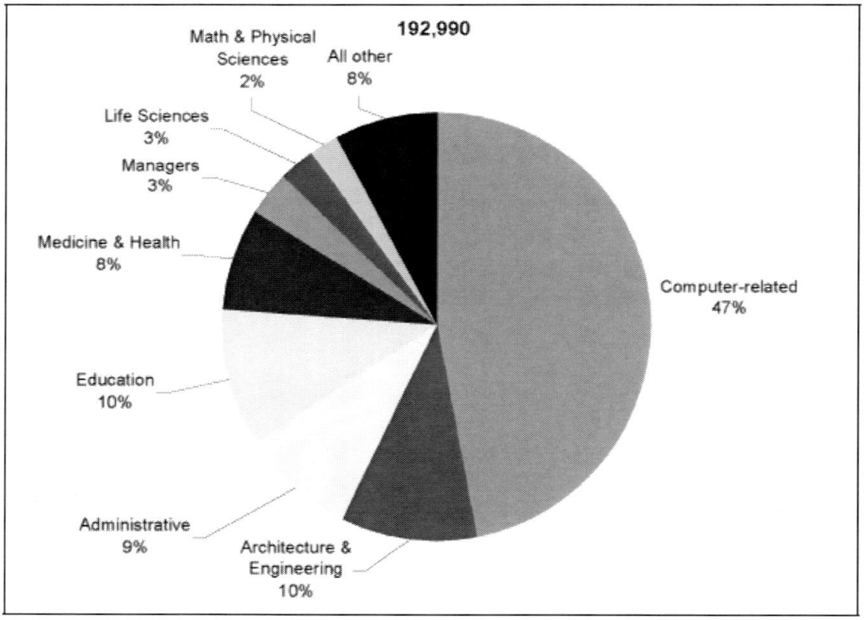

Source: CRS presentation of data from U.S. Citizenship and Immigration Services.

Note: H-1B beneficiaries include workers renewing their visas as well newly arriving workers.

Figure 5. Occupations of H-1B Worker Beneficiaries in FY2010.

In terms of the broad trade group categories in which the H-1B beneficiaries were employed in FY2010, the leading trade group was computer systems design, which employed about a third (34%) of the H-1B workers (*Figure 6*). Colleges and universities employed 10% of the H-1B beneficiaries. Similar numbers of H-1B beneficiaries were employed in the following trade/industry sectors: management and scientific consulting; architecture and engineering; hospitals; and, securities, commodities, and brokerages. Each of these groups have hired about 3% of the H-1B beneficiaries. *Figure 6* shows that a noteworthy portion of the H-1B beneficiaries (38%) were not employed in one of the top trade/industry sectors, suggesting they are employed across the various trade/industry sectors.[21]

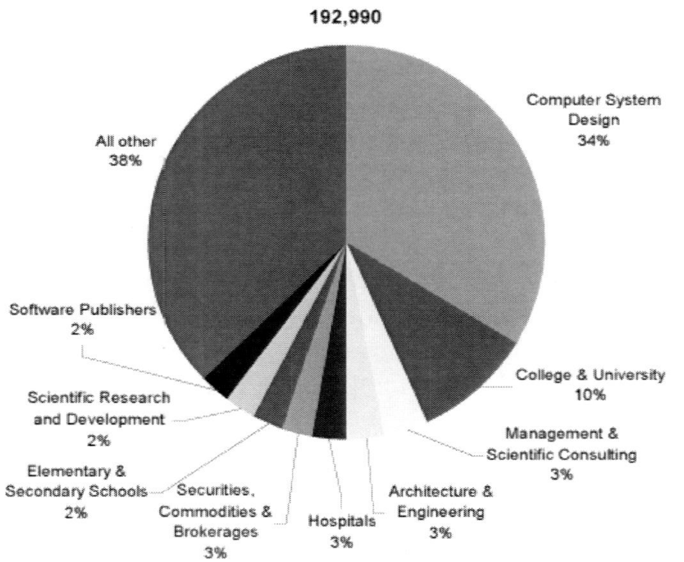

Source: CRS presentation of data from U.S. Citizenship and Immigration Services.
Note: H-1B beneficiaries include workers renewing their visas as well newly arriving workers.

Figure 6. Top Trade/Industry Sectors Hiring H-1B Worker Beneficiaries, 2010.

Other Professional Specialty Workers: TN and E-3 Visas

There are two nonimmigrant visa categories quite similar to H-1B visas that are designated for temporary professional workers from specific countries. These visas are based upon specific trade agreements foreign nations have signed with the United States. Canada and Mexico temporary professional workers may enter according to terms set by the North American Free Trade Agreement (NAFTA) on TN visas.

The E-3 treaty professional visa is a temporary work visa limited to citizens of Australia.[22] Occupationally, they mirror the H-1B visa in that the foreign worker on an E-3 visa or a TN visa must be employed in a specialty occupation. They do not, however, require a labor attestation.

There are several other employment-based nonimmigrant visas for which foreign students with STEM degrees may be eligible. These nonimmigrants include cultural exchange workers on J visas; multinational executive and specialist employees on L visas; international investors on E visas; and persons with outstanding and extraordinary ability on O visas. For further information, see the *Appendix*, "Other High-Skilled Temporary Employment Categories."

PATHWAYS TO LEGAL PERMANENT RESIDENCE

The H-1B visa and OPT often provide the link for foreign students to become employment-based LPRs. Many anecdotal accounts tell of foreign students who are hired by U.S. firms as they are completing their programs. The employers obtain H-1B visas for the recent graduates or hire them as OPT to extend their F-1 visas.

According to the DHS: "This extension of the OPT period for STEM degree holders gives U.S. employers two chances to recruit these highly desirable graduates through the H-1B process, as the extension is long enough to allow for H-1B petitions to be filed in two successive fiscal years."[23] If the temporary foreign workers meet expectations, the employers may also petition for them to become LPRs through one of the employment-based immigration categories.[24]

Over 90% of 140,000 employment-based LPRs annually are adjusting to LPR status within the United States rather than newly arriving from abroad. Presumably, many of these foreign nationals had originally entered the United States as foreign students or temporary workers.

Most foreign nationals seeking to qualify for a nonimmigrant visa, however, must demonstrate that they are not coming to reside permanently. Temporary workers who are H-1B visa holders are permitted to petition for a LPR visa at the same time that they file for an H-1B visa, a policy exception known as dual-intent.[25]

Permanent Employment-based Admissions

Admissions and adjustments for LPR status are subject to a complex set of numerical limits and preference categories that give priority for admission on the basis of family relationships, needed skills, and geographic diversity. The INA establishes an annual worldwide level of 140,000 employment-based preference immigrants, and that ceiling includes the accompanying spouse and children in addition to the principal (i.e., qualifying) alien.[26] The employment-based preference categories are as follows:

- *first preference*: 40,040 for priority workers who are persons of extraordinary ability in the arts, sciences, education, business, or athletics; outstanding professors and researchers; and certain multinational executives and managers;
- *second preference*: 40,040 for members of the professions holding advanced degrees or persons of exceptional ability;
- *third preference*: 40,040 skilled workers with at least two years training, professionals with baccalaureate degrees, and unskilled workers in occupations in which U.S. workers are in short supply;
- *fourth preference*: 10,000 for special immigrants who largely consist of religious workers, certain former employees of the U.S. government, and undocumented juveniles who become wards of the court; and
- *fifth preference*: 10,000 for investors who invest at least $1 million (or less money in rural areas or areas of high unemployment) to create at least 10 new jobs.[27]

The employment-based visas made up 14.2% of the total one million LPRs in FY2010.[28]

Foreign nationals with STEM degrees may qualify under several of the preference categories, depending on their talent, educational attainment, expertise, and experience. Employers who seek to hire prospective employment-based immigrants through the second and third preference categories also must petition DOL on behalf of the alien. The prospective immigrant must demonstrate that he or she meets the qualifications for the particular job as well as the preference category. If DOL determines that a labor shortage exists in the occupation for which the petition is filed, labor certification will be issued. If there is not a labor shortage in the given occupation, the employer must submit evidence of extensive recruitment efforts in order to obtain certification.[29]

From FY2000 through FY2009, 145,390 foreign nationals qualified as first preference principals, i.e., priority workers who are persons of extraordinary ability, outstanding professors and researchers or certain multinational executives and managers. Of these, 21% reported they were engaged in STEM occupations, using the broader NSF definition. As *Figure 7* depicts, almost half (49%) of the first preference principals reported executive, administrative or managerial professions.[30]

As *Figure 7* illustrates, 61% of the second preference principals reported they were engaged in STEM occupations during FY2000-FY2009. The broader NSF definition of STEM fields was used for this analysis. In other words, STEM occupations dominated among the 184,382 foreign nationals who were admitted over this decade as members of the professions holding advanced degrees or persons of exceptional ability.

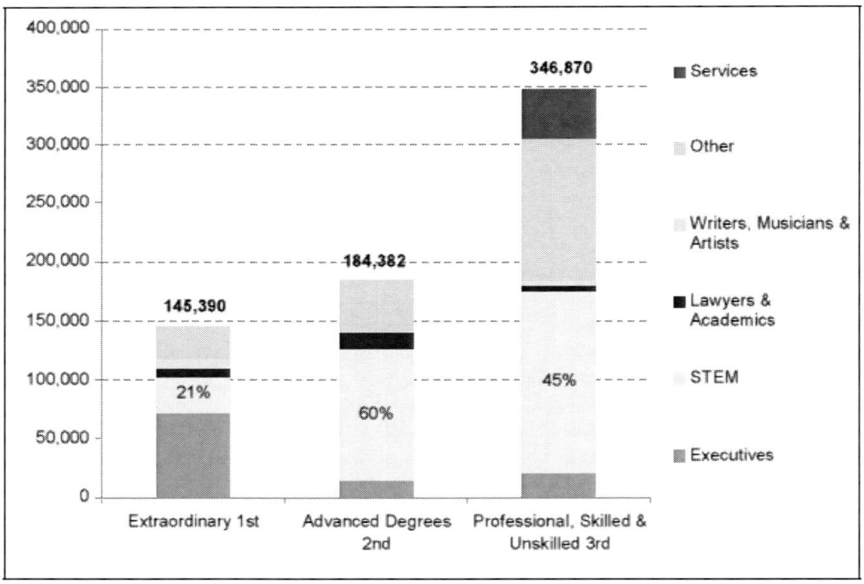

Source: CRS analysis of data from the DHS Office of Immigration Statistics and the former Immigration and Naturalization Service.

Notes: STEM includes biological, agricultural, and environmental life sciences; computer and information sciences; mathematics and statistics; the physical sciences; psychology; the social sciences; engineering; and health fields.

Figure 7. Occupations of 1st through 3rd Preference Employment-based Principals, FY2000-FY2009.

STEM occupations also made up a plurality of the third preference principals (i.e., skilled workers with at least two years training, professionals with baccalaureate degrees, or unskilled workers in occupations in which U.S. workers are in short supply). From FY2000 through FY2009, 45% of the 346,870 foreign nationals admitted as third preference principals reported they worked in STEM occupations.

In total, foreign nationals reporting STEM occupations made up 44% of all of the 676,642 LPRs who were employment-based principals in the first, second and third preference categories during the decade of FY2000-FY2009. Of all of the LPRs reporting STEM occupations (297,668) over this decade, 52% entered as third preference principals.

Employment-Based I-485 Applications Pending

The volume of employment-based immigration cases pending is an important factor in the STEM debate. The USCIS maintains a system of approved employment-based I-485 applications (i.e., the Application to Register Permanent Residence or Adjust Status) that are pending, which provides another source of data on the number of approved employment-based LPRs. Known as the I-485 Inventory, these data are available by preference category and by top countries. These I485 data include the employment-based applicants who plan to adjust their immigration status within the United States. The prospective employment-based LPRs who would be new arrivals from abroad are not included in the I-485 inventory, because they would not need to file I-485 petitions.

As *Figure 8* shows, STEM graduates seeking LPR status are likely to wait in line to obtain LPR status, but those who meet the criteria of the extraordinary ability or advanced degrees preference categories have a much short wait. That there are no third preference I-485 applications filed after August 2007 in the I-485 Inventory suggests that USCIS has not been approving many since the 2007 visa retrogression pushed back the visa priority dates. These data likely understate the number of cases pending.[31]

LEGISLATIVE HISTORY OF STEM VISAS

Legislative proposals for STEM visas emerged during the debate over Comprehensive Immigration Reform in the mid-2000s. Various options aimed at retaining STEM graduates of American universities were folded into the legislation that was moving at the time. The approach was two-fold: (1) foster the recruitment and retention of STEM foreign students; and (2) enable foreign nationals with graduate degrees in STEM fields to adjust to LPR status without waiting in the queue of numerically-limited visas. Although the ideas for STEM visas continue to be discussed, legislative action on STEM visas stalled along with the broader Comprehensive Immigration Reform efforts in 2007.

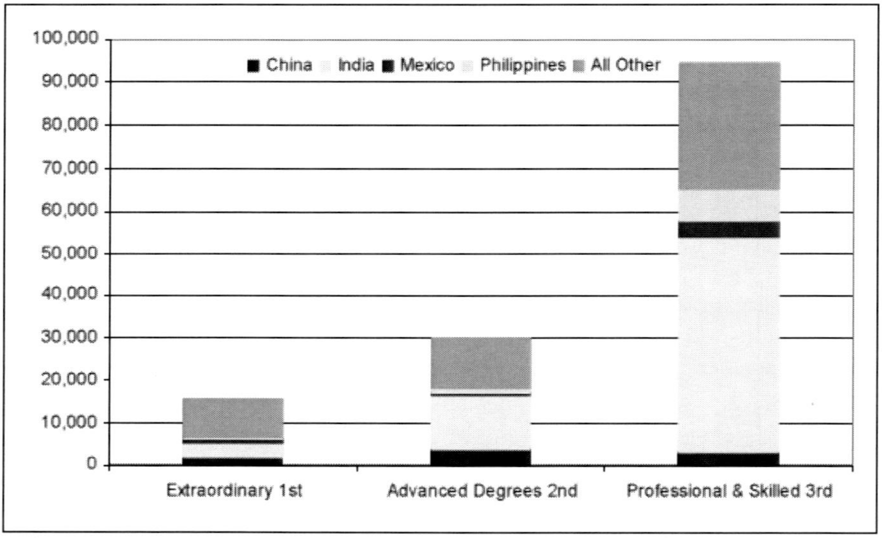

Source: USCIS I-485 Inventory of pending cases, as of January 5, 2012.

Note: There were 140,245 cases; however, no "Professional and Skilled" (3rd) preference applications filed after August 2007 are included in the I-485 Inventory thus far.

Figure 8. Employment-Based I-485 Applications Pending January 2012 by Preference and Top Countries.

During the 109th Congress, the Senate-passed Comprehensive Immigration Reform legislation (S. 2611) was the first piece of legislation with STEM visa provisions that received action. The legislation would have exempted from numerical limits foreign nationals who had worked in the United States for three years and who had earned an advanced degree in a STEM field. More specifically, §507 of S. 2611 as passed by the Senate would have permitted foreign students on a proposed STEM student visa to adjust to LPR status. Also, §508 would have allowed an unlimited number of foreign nationals who had earned an advanced degree in a STEM field and had been working in a related field in the United States during the three years preceding to become LPRs. The 109th House of Representatives did not pass comparable legislation.

In the 110th Congress, S.Amdt. 1150 to S. 1348, the Comprehensive Immigration Reform legislation that failed a cloture vote in the Senate, as well as H.R. 1645 (Security Through Regularized Immigration and a Vibrant Economy Act of 2007 or the STRIVE Act), similarly would have created STEM visas. The legislative proposals in the 110th Congress would have established a STEM foreign student visa category and would have permitted an unlimited number of foreign nationals who had earned STEM degrees and had been working in a related field in the United States during the three years preceding to become LPRs. Foreign students seeking STEM visas would not have needed to demonstrate an intent of departing the United States upon completion of their studies. Students in this category would have also been permitted to pursue Optional Practical Training for periods of up to 24 months after completing their degree. Employers would have been required to pay foreign students the higher of the average or prevailing wage in the field of employment.

SELECTED LEGISLATION IN THE 112TH CONGRESS

There may be renewed interest in establishing STEM visas in the 112th Congress, and several bills have been introduced in both chambers. The House Committee on the Judiciary held two hearings on these issues in 2011.[32] These issues also arose during the Senate Committee on the Judiciary hearing on the economic rationale for immigration reform.[33] No legislation on STEM visas, however, has moved through committees as of the date of this report.[34]

The Immigration Driving Entrepreneurship in America Act of 2011 or the IDEA Act of 2011 (H.R. 2161), the American Innovation and Education Act of 2011 (H.R. 3146), and the Science, Technology, Engineering, and Mathematics Visa Act of 2011 or STEM Visa Act of 2011 (S. 1986) while not identical, do include immigration provisions that are comparable in terms of STEM visas. The bills would amend the INA to establish a priority worker immigrant visa for an alien who has a qualifying: (1) master's or higher degree in a field of science, technology, engineering, or mathematics (STEM degree) from a U.S. institution of higher education; and (2) employment offer from a U.S. employer. S. 1986, H.R. 2161 and H.R. 3146 also would eliminate the foreign residency requirement for certain foreign students, enabling them to have dual intent.[35] The bills further would allow extensions of stay based upon a lengthy labor certification or priority adjudication for F-1 students, H-1B workers, and L intracompany transfers.

S. 1986, H.R. 2161 and H.R. 3146 would also amend the INA to impose a fee of $2,000 per petition on an employer who files an application to hire a foreign worker. Certain not-for-profit and university-based employers would be exempt from paying the fee. Funds from the fees collected would be deposited in the STEM Education and Training Account, which the bills would establish in the Treasury. S. 1986, H.R. 2161 and H.R. 3146 furthermore would require that 60% of the funds be used for a STEM scholarship program for low-income U.S. students enrolled in degree programs in science, technology, engineering, or mathematics.[36]

The Stopping Trained in America Ph.D.s From Leaving the Economy Act of 2011 or the STAPLE Act (H.R. 399) would amend the INA to exempt from: (1) direct numerical limitations aliens who have earned a Ph.D. degree in a STEM field from a U.S. institution of higher education and who have an offer of employment from a U.S. employer in a field related to such degree; and (2) H-1B visa numerical limitations aliens who have earned a Ph.D. degree in a STEM field from a U.S. institution, so long as that degree is required for the job the petitioning employer is filling.

The Startup Act of 2011[37] (S. 1965) would create a new conditional resident visa for up to 50,000 foreign students who have a master's or higher degree in a STEM field from a U.S. institution of higher education. The recipients of these new STEM visas would be granted conditional residence status contingent upon their remaining actively engaged in a STEM field for five consecutive years, after which they would be able to adjust to legal permanent resident status.

POLICY DISCUSSION

The fundamental policy question for Congress is: should the United States create additional pathways for STEM graduates to remain in the United States permanently? Some policy makers consider establishment of STEM visas to be a natural and positive chain of events, arguing that it would be inadvisable to educate talented young people only to make them leave to work for foreign competitors. Others maintain employment-based LPR admissions are highly skewed toward high-skilled immigrants, which already provide for generous flows of foreign nationals with STEM degrees. A corollary factor is that pulling prospective LPRs with graduate STEM degrees out of the numerically-limited employment-based categories would free up visas for the other prospective LPRs waiting in the employment-based queue. The policy discussion opens with two competing perspectives on the immigration prospects of international students with STEM degrees. Four themes of debate conclude the report: assessments of U.S. labor market needs; the competing definitions of STEM fields; the role of colleges and universities as gatekeepers; and the context of broader immigration reforms.

Two Perspectives on STEM Immigration Prospects

At the crux of the legislative debate is whether expedited immigration pathways for STEM foreign graduate students would be in the national interest, and two widely-cited studies of the issue are illustrative of this debate. The two studies featured below yield divergent findings, in large part because they use different research designs and study different STEM populations. This section does not purport to say that one research design is superior to another or that one target population is more meritorious than the others. Rather, it offers two examples with distinct approaches to highlight the competing perspectives. One approach is from the vantage point of the immigration caseloads in the visa categories that STEM degree holders currently use to become LPR; it explores the question of how long would foreign nationals from China and India who have STEM degrees wait to become LPRs of the United States. The other approach is a retrospective analysis of the foreign nationals who earn doctorates in STEM fields; it addresses the issue of retention rates of foreign STEM doctorates in the United States 10 years after graduation.

STEM Graduates Face Long Waits for LPR Status

A 2011 study by Stuart Anderson of the National Foundation for American Policy (NFAP) examined prospective LPRs with approved petitions to immigrate who were waiting in the queue for a numerically limited visa. Since the official data sources do not provide a complete total of cases in the pipeline, NFAP estimated some of the missing components.[38] The NFAP analysis of the data indicate that the majority of employer-sponsored LPRs at the end of 2010 were from India and China.

NFAP's Anderson estimated that a professional worker from India who applies for a professional and skilled (third preference) visa would wait approximately 70 years to obtain LPR status.[39] *Figure 9* presents the NFAP estimates for the upper end, i.e., maximum number of years, that prospective LPRs have been waiting and might continue to wait for an LPR visa according to when they filed their petitions.

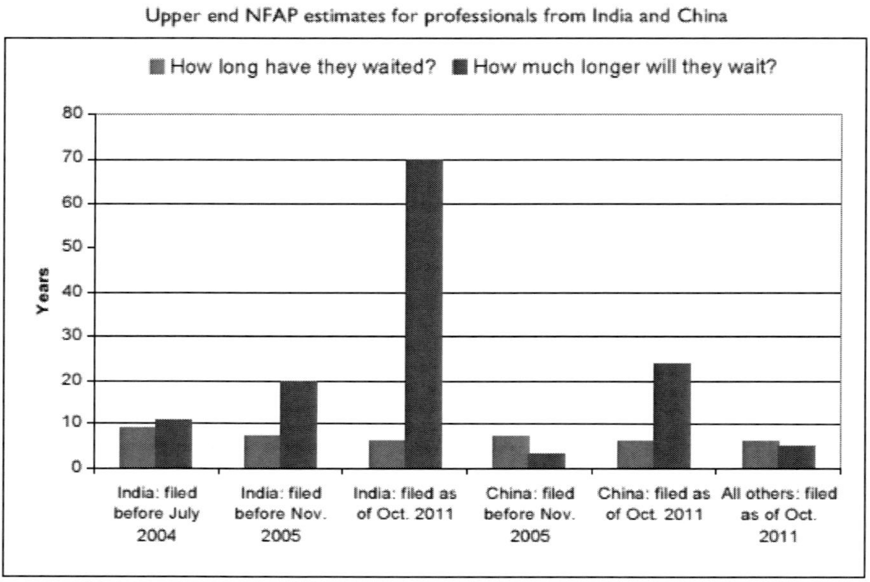

Source: Waiting and More Waiting: America's Family and Employment-based Immigration System, Table 2, National Foundation for American Policy (NFAP), October, 2011.

Notes: Bars depict the upper end (i.e., maximum number of years) of the NFAP estimates.

Figure 9. Projected Wait Times for Third Preference LPRs.

The NFAP study of the pending caseload also estimated that a professional worker from China who applies for a third preference visa would wait approximately 24 years to obtain LPR status.[40] As *Figure 9* shows, the NFAP projected further that countries other than India and China would have wait times for a third preference of four to six years. Anderson concluded "America would lose much talent as U.S.-based businesses would need to hire or place such skilled individuals abroad, rather than invest in a green card process likely to last decades."[41]

The NFAP study indicates that foreign nationals holding advanced degrees (second preference) face far shorter wait times. Only India and China are mentioned as having a noteworthy number of second preference cases pending. Indians and Chinese seeking second preference LPR visas today would have waits of approximately six years, according to the NFAP study.[42]

NFAP's Anderson estimates that an exemption of 50,000 visas a year for foreign students with STEM graduate degrees would eliminate the backlog of cases for those holding advanced degrees (second preference) within two years. He further projects that it would ease the backlog of professional and skilled (third preference) workers, eliminating the backlog within 10 years.[43]

Stay Rates of STEM Graduates Remain High

On-going research that Michael G. Finn has conducted for the National Science Foundation (NSF) on foreign nationals receiving science and engineering doctorates 10 years after graduation offers a different perspective. According to his latest published analysis, the 2009 stay rate for all foreign doctorate recipients was 64% for those graduating five years earlier and 66% for those graduating 10 years earlier. *Figure 10* has been reproduced from his report.[44]

Finn found that the 2009 stay rate of doctorate recipients on temporary resident visas at the time of graduation behaved slightly differently for different cohorts. For those graduating five years earlier, the stay rate was down slightly from that recorded two years ago. However, for those graduating 10 years earlier, the stay rate in 2009 reached an all-time high. Finn concluded that stay rates for temporary foreign nationals receiving science and engineering doctorates overall have never been higher.[45]

Finn also studied prospective immigrants from India and China, and he reached this conclusion:

> China and India are countries of special interest because they account for a large and growing share of new doctorate recipients and are subject to some restrictions not faced by most other countries when seeking permanent resident status. When comparing the history of stay rates for these two countries with that of all other countries, however, there is no apparent evidence that visa restrictions have reduced stay rates for China and India.46

At first glance it seems difficult to reconcile Finn's conclusion about prospective LPRs from China and India, with Anderson's estimates of how long prospective LPRs from China and India might wait before visas would be available for them. From Anderson's perspective the wait times are too long for prospective employment-based immigrants in the professional and skilled category. From Finn's perspective, those with science and engineering doctorates have sufficient time and opportunity to become LPRs through one of the extraordinary ability or advanced degrees categories.

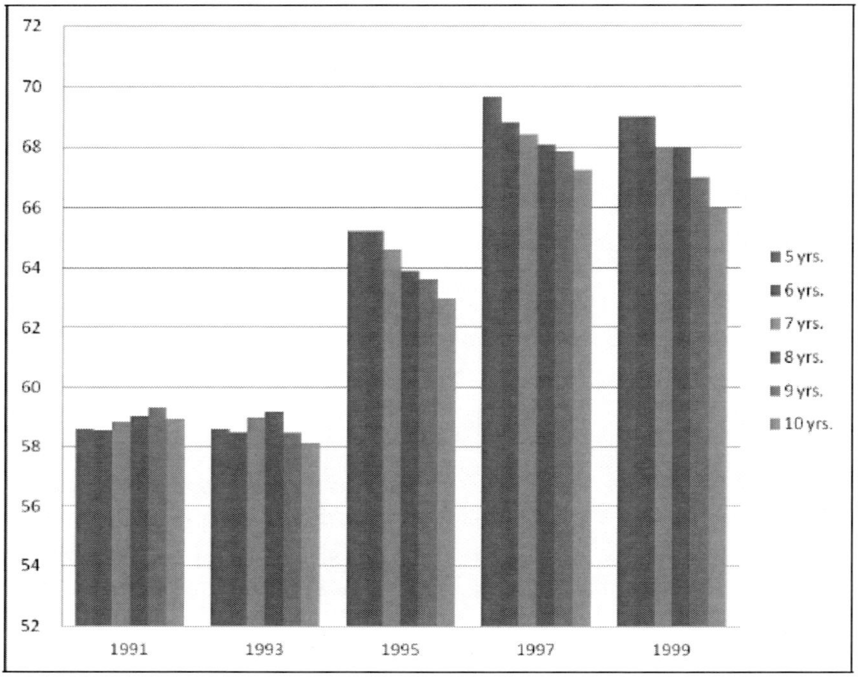

Source: reproduced from Stay Rates of Foreign Doctorate Recipients from U.S. Universities, 2009, by Michael G. Finn, January 2012.

Figure 10. Percentage of Foreign Nationals Receiving Science and Engineering Doctorates Who Were in the United States Five to 10 Years After Receipt of Doctorate, for Doctorates Awarded in 1991, 1993, 1995, 1997, and 1999.

U.S. Labor Market Needs

A broad consensus of business, academic, and policy leaders warn that the United States is on the verge of STEM workforce shortages, which will diminish U.S. global economic competitiveness. Some analysts warn that without retaining more STEM graduates, the United States would suffer a loss of entrepreneurship, would decline in the knowledge economy, and would lose its premier place in the world of innovation. Proponents of STEM visas cite the substantial contributions of foreign-born residents in the United States to international patent creation (25.6% in 2006). “If we don't keep the skilled people in this country after they are educated in our universities and our institutions," U.S. Chamber of Commerce President Tom Donohue said, "companies have a simple choice: If we can't get them here and they go somewhere else, we send the work to where they are." The perspective of many in the business community is that skilled immigrants make jobs rather than take jobs.[47]

Some researchers maintain that creating a STEM visa would not be prudent at this time, given the soft labor market for some STEM occupations. To support this argument, they point out that the unemployment rate of 5.2% for computer and mathematical occupations was greater than the 4.7% unemployment rate of college graduates generally in 2010, and that the unemployment rate of chemists and material scientists reached its highest level in 40 years in 2011 (6.1%).[48] Earlier research from 2007 found that STEM shortages were either overstated or misunderstood. One scholar testified about research indicating that some employers already use high-skilled temporary visa categories to recruit foreign workers at below market rates. This researcher also argued that employers have undue control over the temporary foreign workers because their legal status hinges on their employment. One study found that universities in the United States actually graduate many more STEM students than are hired each year. In 2011, a policy researcher testified that the science and engineering labor force has a substantial supply and that it is best characterized as a loose (not tight) labor market.[49]

Defining/Refining STEM Fields

As noted at the onset, there is no consensus on the definition of STEM fields within academia or federal agencies. Concerns have been expressed that linking LPR status to a STEM degree without a clearly accepted definition might well have unintended consequences.

For example, as a result of revisions in federal Standard Occupational Classification (SOC) in STEM occupations, there was a significant reclassification of employees. Roughly 5% of jobs in the U.S. labor force currently are considered STEM positions. The broader the STEM fields are defined, the more likely that no critical discipline would be omitted; however, a broad definition, some maintain, would also be more susceptible to gaming and abuses.[50]

Others point out that even the narrowest definition would include widely heterogeneous disciplines with widely varied labor market conditions. For example, the engineering profession, which has evolved and splintered into sub-specialties over time, was defined in 1993 as follows:

> Civil engineers had more specialized training as structural engineers, dam engineers, waterpower engineers, bridge engineers; mechanical engineers as machine-design engineers, industrial engineers, motive-power engineers; electrical engineers as power and communication engineers (and the latter divided eventually into telegraph, telephone, radio, television, and radar engineers, whereas the power engineers divided into fossil-fuel and nuclear engineers); mining engineers as metallic-ore mining engineers and fossil-fuel mining engineers (the latter divided into coal and petroleum engineers).51

Today, engineering also encompasses such professions as computer engineers designing microchips that use light pulses and biomedical engineers researching the structure of human cells to foster tissue growth. To deal with the issue of defining STEM fields, one witness during a 2011 House hearing recommended first clarifying the policy motivations for the immigration benefit and then assigning the task of enumerating eligible fields to an agency or inter-agency work-group as an option.[52]

Confounding the discussion of what disciplines should be considered "STEM fields" is the perspective that STEM degrees are not as important as competency in a core set of STEM-based capabilities (referred to as STEM competency). There is an increasing demand for STEM competency in non-STEM occupations, which enables many people with STEM competencies to choose higher-paying career options in other fields. One report finds that out of every 100 students with a bachelor's degree, 19 graduate with a STEM degree but only eight are working in STEM occupations ten years after graduation.[53]

Colleges and Universities as Immigration Gatekeepers

Proponents of STEM visa legislation assert that completion of a graduate degree in a STEM field at a U.S. institution is an excellent criterion for selecting LPRs. They argue that the United States should not be sending back the world's brightest students after educating them. They maintain that current policy not only fosters a reverse brain drain from our economy, but it also equips nations who compete with the United States in the global economy with graduates trained by our world-class universities. A witness at a 2011 hearing on STEM visas testified that a significant majority of foreign students—85% of Indian and Chinese students and 72% of European students—reported concerns about obtaining work visas. The witness warned that theUnited States was falling behind in the global competition for talent.[54]

Other policy makers question the potential consequences of linking LPR status to student visas. For example, the Chairman of the House Committee on the Judiciary warned that establishment of STEM visas would create an incentive for some schools to recruit tuition-paying foreign students with the lure of LPR status upon graduation and cited reports from Australia, where some institutions of higher education were deemed to be "selling education for visas." Another study of the unintended consequences of Australia's linkage of its skilled migration program with international students reported " the failure of some former overseas students to achieve employment outcomes that were commensurate with their qualifications; and failure to obtain skill levels that would meet Australia's skill needs." Further concerns are the possibilities that the incentive of STEM visas may lead to foreign nationals displacing qualified American students or may lead to an over-production of advanced STEM degrees. Some experts advise that institutional eligibility should be carefully defined to prevent "diploma mills" from cashing in on the STEM visas and offer options such as limiting the eligible institutions to those requiring residency at the institution or receiving research funding from NSF or the National Institutes of Health.[55]

Context of Broader Immigration Reform

The challenge in formulating an immigration system is structuring it to represent the country's values, priorities and needs. The difficulty is designing a system that operationalizes these specific values, priorities, and needs into immigration policy. STEM visas are appealing to some because they offer an option that appears to be solely merit-based. Yet, such specification of merit itself provokes debate over what human capital, personal traits, and values prospective immigrants should bring to a country. Some maintain that the STEM visas legislation has broad enough support to be enacted without tackling the thornier issues that confound comprehensive immigration reform, such as the immigration status of unauthorized students who were brought to the United States as children by their parents.[56] Others maintain it merely begs the question by increasing legal immigration without addressing wider reforms.

Appendix. Other High-Skilled Temporary Employment Categories

Cultural Exchange Workers: J Visas[57]

The broadest category for cultural exchange is the J visa, which includes professors and research scholars, students, foreign medical graduates, teachers, resort workers, camp counselors and au pairs who are participating in an approved exchange visitor program. U.S. Department of State's Bureau of Educational and Cultural Affairs (BECA) is responsible for approving the cultural exchange programs. J visa holders are admitted for the period of the program. Many foreign nationals on J-1 visas are permitted to work as part of their cultural exchange program participation. The J cultural exchange visas have expanded over time from visas issued for educational, research or scholarship purposes to visas issued for programs engaged in more mundane tasks, such as child care, resort work, or camp counseling. Today, the J visas may be issued to over a dozen subcategories of exchange visitors. Many of the J visa holders are coming to work in the United States, and some may be employed in STEM fields.

Multinational Executive and Specialist Employees: L Visas

Intracompany transferees who are executive, managerial, or have specialized knowledge, and who are employed with an international firm or corporation are admitted on the L visas. The prospective L nonimmigrant must demonstrate that he or she meets the qualifications for the particular job as well as the visa category. The foreign national must have been employed by the firm for at least six months in the preceding three years in the capacity for which the transfer is sought. The alien must be employed in an executive capacity, a managerial capacity, or have specialized knowledge of the firm's product to be eligible for the L visa. STEM graduates are most likely to fulfill the specialized knowledge element. The INA does not require firms who wish to bring L intracompany transfers into the United States to demonstrate that U.S. workers will not be adversely affected in order to obtain a visa for the transferring employee.

International Investors: E Visas

Aliens who are treaty traders enter on E-1 visas, whereas those who are treaty investors enter on E-2 visas. An E-1 treaty trader visa allows a foreign national to enter the United States for the purpose of conducting "substantial trade" between the United States and the country of which the person is a citizen. An E-2 treaty investor can be any person who comes to the United States to develop and direct the operations of an enterprise in which he or she has invested, or is in the process of investing, a "substantial amount of capital." Entrepreneurial STEM graduates, in particular, may utilize the E-2 visa. Both these E-class visas require that a treaty exist between the United States and the principal foreign national's country of citizenship.[58]

Persons with Outstanding and Extraordinary Ability: O Visas

Persons with extraordinary ability in the sciences, arts, education, business, or athletics can be admitted on O visas. Generally, the O visa is reserved for the highest level of accomplishment and covers a fairly broad set of occupations and endeavors, including athletics and entertainment. Regulations implementing the O-1 visa define extraordinary ability in the field of science, education, business, or athletics as "a level of expertise indicating that the person is one of the small percentage who have arisen to the very top of the field of endeavor."[59] In FY2010, the State Department issued 8,589 O-1 visas.

End Notes

[1] For a fuller discussion of science, technology, engineering, or mathematics (STEM) fields, see CRS Report R42430, America COMPETES 2010 and the FY2013 Budget, by Heather B. Gonzalez; and CRS Report R42470, An Analysis of STEM Education Funding at the NSF: Trends and Policy Discussion, by Heather B. Gonzalez.

[2] For a complete discussion of the employment-based visa queue, see CRS Report R42048, Numerical Limits on Employment-Based Immigration: Analysis of the Per-Country Ceilings, by Ruth Ellen Wasem.

[3] The other two economic preference categories are: special immigrants who largely consist of religious workers and certain former employees of the U.S. government; and investors who invest at least $1 million (or less money in rural areas or areas of high unemployment) to create at least 10 new jobs. §203(b) of INA; 8 U.S.C. §1153.

[4] Committee on Science, Engineering, and Public Policy, Policy Implications of International Graduate Students and Postdoctoral Scholars in the United States, National Academies Press, 2005; Vivek Wadhwa, Guillermina Jasso, and Ben Rissing, et al., Intellectual Property, the Immigration Backlog, and a Reverse Brain-Drain, part III, Duke University, New York University, Harvard Law School and the Ewing Marion Kauffman Foundation, August 2007; and, David Langdon, George McKittrick, and David Beede, et al., STEM: Good Jobs Now and for the Future, U.S. Department of Commerce Economics and Statistics Administration, 03-11, July 2011, http://www.esa.doc.gov/sites/ default/ files /reports/documents/stemfinalyjuly14_1.pdf.

[5] Thomas Lemieux, "The Changing Nature of Wage Inequality," Journal of Population Economics, vol. 21, no. 1 (January 2008); David Card and John E. DiNardo, "Skill-Biased Technological Change and Rising Wage Inequality: Some Problems and Puzzles," Journal of Labor Economic, vol. 20, no. 4 (2002); B. Lindsay Lowell, Hal Salzman, and Hamutal Bernstein, Steady as She Goes? Three Generations of Students through the Science and Engineering Pipeline, Rutgers University and the Institute for the Study of International Migration, October 2009; and, U.S. Congress, House Committee on the Judiciary, Subcommittee on Immigration Policy and Enforcement, STEM the Tide: Should America Try to Prevent an Exodus of Foreign Graduates of U.S. Universities with Advanced Science Degrees, testimony of Dr. B. Lindsey Lowell, 112th Cong., 1st sess., October 5, 2011; and David Langdon, George McKittrick, and David Beede, et al., STEM: Good Jobs Now and for the Future, U.S. Department of Commerce Economics and Statistics Administration, 03-11, July 2011, http://www.esa.doc.gov/sites/default/files/reports/documents/ stemfinaly july14_1.pdf.

[6] Peter Einaudi, Two Decades of Increasing Diversity More than Doubled the Number of Minority Graduate Students in Science and Engineering, National Science Foundation, NSF 11-319, July 2011, http://www.nsf.gov/statistics/infbrief/ nsf11319/nsf11319.pdf.

[7] For a detailed list of the Immigration and Custom Enforcement STEM-designated degree programs, see http://www.ice.gov/doclib/sevis/pdf/stem-list-2011.pdf.

[8] National Center for Education Statistics, Postsecondary Awards in Science, Technology, Engineering, and Mathematics, by State: 2001 and 2009, U.S. Department of Education, April 2011, http://nces.ed.gov/pubs2011/ 2011226.pdf.

[9] CRS Report RL32188, Monitoring Foreign Students in the United States: The Student and Exchange Visitor Information System (SEVIS), by Alison Siskin; and CRS Report RL31146, Foreign Students in the United States: Policies and Legislation, by Chad C. Haddal.

[10] Nonimmigrants are admitted for a designated period of time and a specific purpose.There are 24 major nonimmigrant visa categories, which are commonly referred to by the letter and numeral that denotes their subsection in the INA; for example, B-2 tourists, E-2 treaty investors, F-1 foreign students, or H-1B temporary professional workers. CRS Report RL31381, U.S. Immigration Policy on Temporary Admissions, by Ruth Ellen Wasem.

[11] Those students who wish to pursue a non-academic (e.g., vocational) course of study apply for an M visa. Their spouses and children may accompany them as M-2 nonimmigrants. Some graduate students use the J visa for cultural exchange, which covers educational, research or scholarship purposes.

[12] 8 C.F.R. 214.2(f)(15)(i).

[13] SEVIS is the foreign student monitoring database. For background on SEVIS, see CRS Report RL32188, Monitoring Foreign Students in the United States: The Student and Exchange Visitor Information System (SEVIS), by Alison Siskin.

[14] Until the 2000s, the enrollment of U.S. citizens in graduate science and engineering programs had not kept pace with that of foreign students in these programs. NSF researchers cite the growth in the numbers of U.S. citizens and permanent residents pursuing graduate-level study in STEM fields in recent years as the reason the percentage of foreign STEM students has remained level. Peter Einaudi, Two Decades of Increasing Diversity More than Doubled the Number of Minority Graduate Students in Science and Engineering, National Science Foundation, NSF 11-319, July 2011, http://www.nsf.gov/statistics/infbrief/ nsf11319/ nsf11319.pdf.

[15] Although based upon data from ICE, the STEM fields presented in Figure 2, Table 1, and Table 2 use the broader NSF definition of STEM. Joan Burrelli, Foreign Science and Engineering Students in the United States , National Science Foundation, NSF 10-324, July 2010, http://www.nsf.gov/statistics/infbrief/nsf10324/..

[16] E-Verify is an electronic employment eligibility verification program that U.S. employers voluntarily use to confirm the new hires' employment authorization through Social Security Administration and, if necessary, DHS databases. CRS Report R40446, Electronic Employment Eligibility Verification, by Andorra Bruno.

[17] 8 C.F.R. 214.2(f)(10).

[18] Attestation was part of a compromise package on H-1B visas that included annual numerical limits in the Immigration Act of 1990 (P.L. 101-649). See CRS Report RL30498, Immigration: Legislative Issues on Nonimmigrant Professional Specialty (H-1B) Workers, by Ruth Ellen Wasem.

[19] INA §212(n); 8 C.F.R. §214.2(h)(4). For a further discussion of labor attestations, see CRS Report RL30498, Immigration: Legislative Issues on Nonimmigrant Professional Specialty (H-1B) Workers, by Ruth Ellen Wasem.

[20] U.S. Citizenship and Immigration Services, Characteristics of H-1B Specialty Occupation Workers, Fiscal Year 2010 Annual Report, Department of Homeland Security, August 4, 2011.

[21] Ibid.

[22] §501 of P.L. 109-13, the Emergency Supplemental Appropriations Act for Defense, the Global War on Terror, and Tsunami Relief, 2005.

[23] U.S. Citizenship and Immigration Services, Extension of Post-Completion Optional Practical Training (OPT) and F1 Status for Eligible Students under the H-1B Cap-Gap Regulations, U.S. Department of Homeland Security, April 2, 2010.

[24] Not all companies, however, seek to convert H-1B employees to LPR status. Research by Professor Ron Hira of the Rochester Institute of Technology indicates that many of the largest users of the H-1B visa sponsor few, if any, of their H-1Bs for permanent residency. U.S. Congress, Senate Committee on the Judiciary, Subcommittee on Immigration, Refugees and Border Security, The Economic Imperative for Enacting Immigration Reform, 112th Cong., 1st sess., July 26, 2011.

[25] The other categories permitted dual intent are intracompany transfers employed with international firms who enter on L visas and foreign nationals with V visas for family-related nonimmigrant. §214(b) of the INA; 8 U.S.C. §1184(b).

[26] Numbers not used in one category roll down to the next category. CRS Report RL32235, U.S. Immigration Policy on Permanent Admissions, by Ruth Ellen Wasem.

[27] For examples of legislation in the 112th Congress that would amend the fifth preference to foster STEM graduates who are entrepreneurs, see H.R. 1114 and S. 565.

[28] Randall Monger and James Yankay, U.S. Legal Permanent Residents: 2010, U.S. Department of Homeland Security Office of Immigration Statistics, Annual Flow Report, March 2011, http://www.dhs.gov/xlibrary/assets/statistics/ publications/lpr_fr_2010.pdf.

[29] See CRS Report RL33977, Immigration of Foreign Workers: Labor Market Tests and Protections, by Ruth Ellen Wasem.

[30] CRS analyzed the U.S. Department of Homeland Security Office of Immigration Statistics LPR data for fiscal years 2000 through 2009. There were a total of 10.3 million LPRs over the decade, of whom 69% reported an occupation. The occupation data analyzed in this report are the subset of 676,642 principal immigrants who received employment 1st, 2nd, or 3rd preference LPR status from fiscal years 2000 through 2009.

[31] A substantial visa retrogression occurred in July 2007. The Visa Bulletin for July 2007 listed the visa priority dates as "current" for the employment-based preferences (except for the unskilled other worker category). On July 2, 2007, however, the State Department issued an Update to July Visa Availability that retrogressed the dates to The State Department offered the following explanation: "The sudden backlog reduction efforts by Citizenship and Immigration Services Offices during the past month have resulted in the use of almost 60,000 Employment numbers.... Effective Monday July 2, 2007 there will be no further authorizations in response to requests for Employment-based preference cases." U.S. Department of State, Bureau of Consular Affairs, Visa Bulletin, No. 108, is available at http://travel.state.gov/visa/frvi/bulletin/bulletin_3266.html.

[32] U.S. Congress, House Committee on the Judiciary, Subcommittee on Immigration Policy and Enforcement, H-1B Visas: Designing a Program To Meet the Needs of the U.S. Economy and U.S. Workers, 112th Cong., 1st sess., March 31, 2011; and U.S. Congress, House Committee on the Judiciary, Subcommittee on Immigration Policy and Enforcement, STEM the Tide: Should America Try to Prevent an Exodus of Foreign Graduates of U.S. Universities with Advanced Science Degrees, 112th Cong., 1st sess., October 5, 2011.

[33] U.S. Congress, Senate Committee on the Judiciary, Subcommittee on Immigration, Refugees and Border Security, The Economic Imperative for Enacting Immigration Reform, 112th Cong., 1st sess., July 26, 2011

[34] For a discussion of H.R. 3012, which would address high-skilled immigration by lifting the employment-based country ceilings, see CRS Report R42048, Numerical Limits on Employment-Based Immigration: Analysis of the Per-Country Ceilings, by Ruth Ellen Wasem.

[35] Dual intent refers to exceptions to §214(b) of the INA that permit foreign nationals to have pending petitions for both nonimmigrant and immigrant visas.

[36] For information on similar fees, see CRS Report RL31973, Programs Funded by the H-1B Visa Education and Training Fee, and Labor Market Conditions for Information Technology (IT) Workers, by Linda Levine and Blake Alan Naughton.

[37] Not to be confused with the StartUp Visa Act of 2011 ((H.R. 1114/S. 565) which would amend the INA to establish an employment-based, conditional immigrant visa (StartUp visa) for a sponsored alien entrepreneur: (1) with required amounts of financial backing from a qualifying investor, government entity, or venture capitalist; and (2) whose commercial activities will generate required levels of employment, revenue, or capital investment.

[38] For a full discussion of the data problems in estimating backlogs of approved petitions, see CRS Report R42048, Numerical Limits on Employment-Based Immigration: Analysis of the Per-Country Ceilings, by Ruth Ellen Wasem.

[39] The study attempted to address the unknown cases pending who were waiting in the queue for a numerically limited visa and estimated that an additional 150,000 prospective LPRs from

India were waiting in the professional and skilled worker preference queue. That estimate yielded a total of 210,000 Indians in the third preference caseload. Stuart Anderson, Waiting and More Waiting: America's Family and Employment-based Immigration System, National Foundation for American Policy, NFAP Policy Brief, October 2011. (Hereinafter Anderson, Waiting and More Waiting: America's Family and Employment-based Immigration System, NFAP Policy Brief, October 2011.)

[40] The study assumed that 55,000 Chinese were already in the third preference caseload. Anderson, Waiting and More Waiting: America's Family and Employment-based Immigration System, NFAP Policy Brief, October 2011.

[41] Anderson, Waiting and More Waiting: America's Family and Employment-based Immigration System, NFAP Policy Brief, October 2011.

[42] Anderson, Waiting and More Waiting: America's Family and Employment-based Immigration System, NFAP Policy Brief, October 2011.

[43] Anderson, Waiting and More Waiting: America's Family and Employment-based Immigration System, NFAP Policy Brief, October 2011.Anderson also factors in the calculation of removing the per-country ceilings on employment-based immigrants. For more on this option, see CRS Report R42048, Numerical Limits on Employment-Based Immigration: Analysis of the Per-Country Ceilings, by Ruth Ellen Wasem.

[44] Michael G. Finn, Stay Rates of Foreign Doctorate Recipients from U.S. Universities, 2009, Division of Science Resources Statistics of the National Science Foundation by Oak Ridge Institute for Science and Education through an interagency agreement with Department of Energy, January 2012, http://orise.orau.gov/files/sep/stay-rates-foreigndoctorate-recipients-2009.pdf. (Hereinafter Finn, Stay Rates of Foreign Doctorate Recipients from U.S. Universities, 2009.)

[45] Finn, Stay Rates of Foreign Doctorate Recipients from U.S. Universities, 2009.

[46] "While stay rates of India and China decline with years since graduation, this decline has been very slight and smaller than the corresponding decline in stay rates observed by all other countries combined." Michael G. Finn, Stay Rates of Foreign Doctorate Recipients from U.S. Universities, 2009, Division of Science Resources Statistics of the National Science Foundation by Oak Ridge Institute for Science and Education through an interagency agreement with Department of Energy, January 2012, http://orise.orau.gov/files/sep/stay-rates-foreign-doctorate-recipients-2009.pdf.

[47] National Academy of Sciences, National Academy of Engineering, and Institute of Medicine, Rising Above the Gathering Storm: Energizing and Employing America for a Brighter Economic Future, Committee on Prospering in the Global Economy of the 21st Century, 2007; Brookings Institution and George Mason University, Immigration Policy: Highly Skilled Workers and U.S. Competitiveness and Innovation, Forum hosted by the Brookings Center for Technology Innovation and the George Mason Center for Science and Technology Policy, February 7, 2011; Vivek Wadhwa, Guillermina Jasso, and Ben Rissing, et al., Intellectual Property, the Immigration Backlog, and a Reverse Brain-Drain, part III, Duke University, New York University, Harvard Law School and the Ewing Marion Kauffman Foundation, August 2007; U.S. Congress, House Committee on the Judiciary, Subcommittee on Immigration Policy and Enforcement, H-1B Visas: Designing a Program To Meet the Needs of the U.S. Economy and U.S. Workers, 112th Cong., 1st sess., March 31, 2011; U.S. Congress, House Committee on the Judiciary, Subcommittee on Immigration Policy and Enforcement, STEM the Tide: Should America Try to Prevent an Exodus of Foreign Graduates of U.S. Universities with Advanced Science Degrees?, 112th Cong., 1st

sess., October 5, 2011; and, Mallie Jane Kim, "Chamber of Commerce, Bloomberg Push Immigration Reform," U.S. News & World Report, September 28, 2011.

[48] U.S. Congress, Senate Committee on the Judiciary, Subcommittee on Immigration, Refugees and Border Security, The Economic Imperative for Enacting Immigration Reform, answers to questions for the record, witness Professor Ron Hira, 112th Cong., 1st sess., July 26, 2011; and Rudy M. Baum, "Unemployment Data Worst In 40 Years," Chemical and Engineering News, March 21, 2012.

[49] Richard Freeman, "The Market for Scientists and Engineers," NBER Reporter, no. 3 (Summer 2007); Rudy M. Baum, "Unemployment Data Worst In 40 Years," Chemical and Engineering News, March 21, 2012; U.S. Congress, Senate Committee on the Judiciary, Subcommittee on Immigration, Refugees and Border Security, The Economic Imperative for Enacting Immigration Reform, answers to questions for the record, witness Professor Ron Hira, 112th Cong., 1st sess., July 26, 2011U.S. Congress, House Committee on the Judiciary, Subcommittee on Immigration Policy and Enforcement, H-1B Visas: Designing a Program To Meet the Needs of the U.S. Economy and U.S. Workers, testimony of Professor Ron Hira, 112th Cong., 1st sess., March 31, 2011; and, U.S. Congress, House Committee on the Judiciary, Subcommittee on Immigration Policy and Enforcement, STEM the Tide: Should America Try to Prevent an Exodus of Foreign Graduates of U.S. Universities with Advanced Science Degrees, testimony of Dr. B. Lindsey Lowell, 112th Cong., 1st sess., October 5, 2011.

[50] "The large decline in the numbers of people who used to be grouped in this field is that the definition of "computer scientists and systems analysts" has become much more precise, and many of the people who might have been included in the more broadly defined profession during the 1990s are now grouped elsewhere under numerous new occupational titles for specialized types of information technology people, such as software engineers, network analysts, and database administrators." Richard Ellis, Effects of Recent Revisions in Federal Standard Occupational Classification (SOC) Categories on Counts of the Employment of STEM Professionals, Commission on Professionals in Science and Technology, May 2, 2007, http://www.cpst.org/NewPubs.cfm; Robert D. Atkinson, "Why the Current Education Reform Strategy Won't Work," Issues in Science and Technology, Spring 2012, pp. 29-36; and U.S. Congress, House Committee on the Judiciary, Subcommittee on Immigration Policy and Enforcement, STEM the Tide: Should America Try to Prevent an Exodus of Foreign Graduates of U.S. Universities with Advanced Science Degrees?, testimony of Barmack Nassirian, 112th Cong., 1st sess., October 5, 2011.

[51] Joseph W. Barker , McGraw-Hill Encyclopedia of Engineering, ed. McGraw-Hill, 2nd ed. (1993), pp. 409-410.

[52] U.S. Congress, House Committee on the Judiciary, Subcommittee on Immigration Policy and Enforcement, STEM the Tide: Should America Try to Prevent an Exodus of Foreign Graduates of U.S. Universities with Advanced Science Degrees?, testimony of Barmack Nassirian, 112th Cong., 1st sess., October 5, 2011.

[53] "At the graduate degree level of educational attainment, it is clear that both Managerial and Professional occupations and Healthcare Professional occupations pay substantially more than STEM." Anthony P. Carnevale, Nicole Smith, and Michelle Melton, STEM: Science, Technology, Engineering, Mathematics, Georgetown University Center on Education and the Workforce, October 2011, V=http://cew.georgetown.edu/stem/.

[54] U.S. Congress, Joint Economic Committee, Fueling Local Economies: Research, Innovation and Jobs, testimony of Robert Litan, 111th Cong., June 29, 2010; U.S. Congress, House Committee on the Judiciary, Subcommittee on Immigration Policy and Enforcement, STEM

the Tide: Should America Try to Prevent an Exodus of Foreign Graduates of U.S. Universities with Advanced Science Degrees?, testimony of Vivek Wadhwa, 112th Cong., 1st sess., October 5, 2011.

[55] U.S. Congress, House Committee on the Judiciary, Subcommittee on Immigration Policy and Enforcement, STEM the Tide: Should America Try to Prevent an Exodus of Foreign Graduates of U.S. Universities with Advanced Science Degrees?, 112th Cong., 1st sess., October 5, 2011; Elsa Koleth, Overseas Students: Immigration Policy Changes 1997– May 2010, Parliament of Australia, Parliamentary Library Background Note, June 18, 2010, http://www.aph.gov.au/ binaries/library/pubs/bn/sp/overseasstudents.pdf.

[56] For a complete analysis of these immigration issues, see CRS Report R41704, Overview of Immigration Issues in the 112th Congress, by Ruth Ellen Wasem; and CRS Report R42036, Immigration Legislation and Issues in the 112th Congress, coordinated by Andorra Bruno.

[57] The Q visa is also an employment-oriented cultural exchange program. Its stated purpose is to provide practical training and employment as well as to share the history, culture, and traditions of the United States and of the exchange visitor's home country. While STEM graduates are not typically recruited in Q exchange programs, they are not precluded.

[58] See CRS Report RL33844, Foreign Investor Visas: Policies and Issues.

[59] 8 C.F.R. §214.2(o)(3)(ii).

In: STEM Workforce Trends … ISBN: 978-1-62257-818-4
Editors: C. Martin Garcia and S. Cook

Chapter 2

STEM: Good Jobs Now and for the Future*

David Langdon, George McKittrick, David Beede, Beethika Khan and Mark Doms

Executive Summary

Science, technology, engineering and mathematics (STEM) workers drive our nation's innovation and competitiveness by generating new ideas, new companies and new industries. However, U.S. businesses frequently voice concerns over the supply and availability of STEM workers. Over the past 10 years, growth in STEM jobs was three times as fast as growth in non-STEM jobs.

STEM workers are also less likely to experience joblessness than their non-STEM counterparts. Science, technology, engineering and mathematics workers play a key role in the sustained growth and stability of the U.S. economy, and are a critical component to helping the U.S. win the future.

- In 2010, there were 7.6 million STEM work- ers in the United States, representing about 1 in 18 workers.

* This is an edited, reformatted and augmented version of the U.S. Department of Commerce, Economics and Statistics Administration Issue Brief #03-11 publication, dated July 2011.

- STEM occupations are projected to grow by 17.0 percent from 2008 to 2018, compared to 9.8 percent growth for non-STEM occupations.
- STEM workers command higher wages, earning 26 percent more than their non- STEM counterparts.
- More than two-thirds of STEM workers have at least a college degree, compared to less than one-third of non-STEM workers.
- STEM degree holders enjoy higher earnings, regardless of whether they work in STEM or non-STEM occupations.

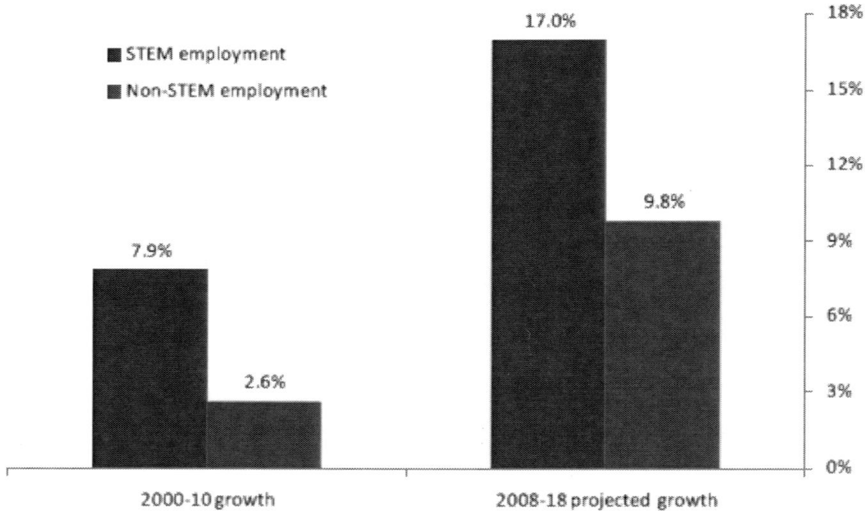

Source: ESA calculations using Current Population Survey public-use microdata and estimates from the Employment Projections Program of the Bureau of Labor Statistics.

Figure 1. Recent and Projected Growth in STEM and Non-STEM Employment.

WHAT IS STEM?

The acronym STEM is fairly specific in nature— referring to science, technology, engineering and math—however, there is no standard definition for what constitutes a STEM job. Science, technology, engineering and math positions consistently make the lists of STEM occupations, but there is less consensus about whether to include other positions such as educators, managers, technicians, health-care professionals or social scientists. In this report, we define STEM jobs to include professional and technical support occupations in the fields of computer science and mathematics, engineering, and life and physical sciences. Three management occupations are also included because of their clear ties to STEM.[1] Because of data limitations, education jobs are not included.[2] Further, we elected not to include social scientists.[3]

Our STEM list contains 50 specific occupation codes (see Appendix Table 1), and in 2010, there were 7.6 million workers in these jobs, or 5.5 percent of the workforce. To better put these jobs into context, we divide STEM occupations into four categories: computer and math, engineering and surveying, physical and life sciences, and STEM managerial occupations.[4] Across all levels of educational attainment, the largest group of STEM jobs is within the computer and math fields, which account for close to half (46 percent) of all STEM employment. Second are engineering and surveying occupations with one-third of all STEM employment, while 13 percent are in the physical and life sciences, and 9 percent in STEM management jobs.

Parallel to our list of STEM occupations, we also identify a set of STEM undergraduate degree fields that span computer science and mathematics, engineering, and life and physical sciences (see Appendix Table 2). We define STEM degree holders as persons whose primary or secondary undergraduate major was in a STEM field. Following similar logic to what we used in our occupation selection, we exclude business, healthcare, and social science majors.

STEM EMPLOYMENT AND WORKER EARNINGS

In 2010, 7.6 million people or 1 in 18 workers held STEM jobs. Although STEM employment currently makes up only a small fraction of total U.S. employment, STEM employment grew rapidly from 2000 to 2010, increasing 7.9 percent. In contrast, employment in non-STEM jobs grew just 2.6 percent over this period (see Figure 1). STEM jobs are projected to grow at a fast pace relative to other occupations. From 2008 to 2018, STEM jobs are expected to grow 17.0 percent compared to just 9.8 percent for non-STEM jobs.[5]

Workers in STEM occupations also earn more on average than their counterparts in other jobs, regardless of their educational attainment. The STEM earnings differential is greatest for those with a high school diploma or less in comparison to their counterparts in a non-STEM field. On average, they earned almost $25 per hour, $9 more per hour than those in other occupations in 2010. It should be noted, however, that only about 1 out of every 10 STEM workers has a high school diploma or less. Those with graduate degrees in a STEM job earned more than $40 per hour, nearly $4.50 more per hour on average than those with non-STEM jobs.

The comparison of wage premiums raises several questions, including to what extent the STEM-earnings premium reflects other characteristics of workers, such as age, and how premiums have evolved over time. A regression analysis – which controls for a variety of demographic, geographic, and other worker characteristics – helps to address these questions. Using Current Population Survey public-use microdata for 1994-2010, we regressed the log of earnings against a standard list of characteristics that have typically been found to be related to earnings including age, marital status, race, ethnicity, region and industry.[6]

After controlling for this set of characteristics, the earning premium diminishes somewhat. However, the fundamental result that STEM workers enjoy large earnings premiums persists, most predominantly for workers with less than a college degree. STEM earnings premiums have also shown persistence over time, and have generally increased since the mid-1990s (see Figure 2). In 2010, workers in STEM jobs with less than a bachelor's degree enjoyed a large premium (more than 30 percent) compared pared with non-STEM workers with the same education level, even after taking other influences on earnings into account.7 The regression-based premiums in 2010 were slightly less for workers with a bachelor's (23 percent) or graduate degree (12 percent), and relatively closer to the premiums found in the simple comparison (without a regression adjustment). The overall regression-based STEM premium was 26 percent in 2010, up from 18 percent in 1994.[8]

Table 1. Average Hourly Earnings of Full-Time Private Wage and Salary Workers in STEM Occupations by Educational Attainment, 2010

	Average hourly earnings		Difference	
	STEM	Non-STEM	Dollars	Percent
High school diploma or less	$24.82	$15.55	$9.27	59.6%
Some college or associate degree	$26.63	$19.02	$7.61	40.0%
Bachelor's degree only	$35.81	$28.27	$7.54	26.7%
Graduate degree	$40.69	$36.22	$4.47	12.3%

Source: ESA calculations using Current Population Survey public-use microdata and estimates from the Employment Projections Program of the Bureau of Labor Statistics.

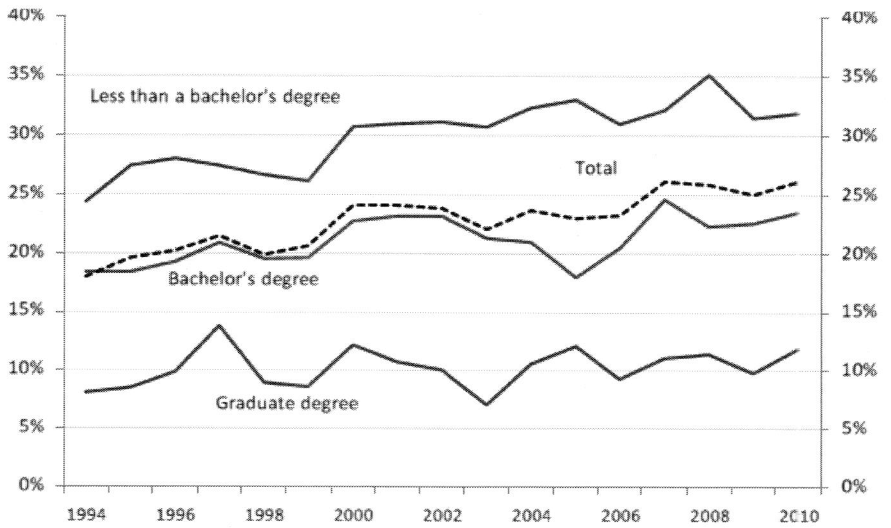

Source: ESA calculations using Current Population Survey public-use microdata files of annual merged outgoing rotation groups from the National Bureau of Economic research.

Note: The estimates are for private wage and salary workers age 25 and over.

Figure 2. Regression-based Hourly Earnings Premiums for STEM Workers 1994—2010.

STEM Jobs and STEM Degrees

The analysis so far has focused on STEM jobs, but conversations about policy most often focus on STEM education. One source of information to analyze the link between STEM jobs and STEM education is the 2009 American Community Survey which collected information on college-educated individuals' undergraduate majors. When examining the relationship between STEM education and STEM jobs, the following patterns emerge.

First, a STEM degree is the typical path to a STEM job, as more than two-thirds of the 4.7 million STEM workers with a college degree has an undergraduate STEM degree. [9] However, this does not necessarily mean that STEM workers' degrees are in the same STEM field as their jobs. For example, only 35 percent of college-educated computer and math workers have a degree in computer science or math while 27 percent majored in the physical or life sciences or engineering. Second, in addition to STEM jobs, STEM degrees also open the door to many other career opportunities.

Almost two-thirds of the 9.3 million workers with a STEM undergraduate degree work in a non-STEM job. Life and physical science majors are the STEM degree holders most likely to work in non-STEM jobs; 81 percent of these graduates work outside the STEM fields. (Note that “non-STEM” occupations include the 28 percent of graduates who work as healthcare practitioners or technicians, and the 12 percent who work in education.) In math, there is a strong pipeline into education jobs, as one-fifth of math majors go on to work in education.

As discussed above, STEM workers earn significantly more than their non-STEM counterparts, but what about the earnings of STEM degree holders who don’t necessarily work in STEM jobs? Using the 2009 American Community Survey public-use microdata, calculations of the regression-adjusted earnings premium of college-educated workers with a STEM degree and/or STEM job showed that all STEM degree holders receive an earnings premium relative to other college graduates, whether or not they end up in a STEM job.

Likewise, college graduates, regardless of their major, enjoy an earnings premium for having a STEM job. The earnings premium for having a STEM job or a STEM degree is quite similar, at 13 percent and 11 percent, respectively. Still, a much larger payoff tends to come when a STEM major goes on to work in a STEM job, as their earnings (all else equal) are about 20 percent higher than those of non-STEM majors working in non-STEM jobs.

Table 2. Employment of Workers Age 25 and Over with a Bachelor’s Degree or Higher, by STEM Occupation and STEM Undergraduate Degree, 2009

Employed persons in thousands

	Total	Total	Computer	STEM degree Math	Engineering	Physical and life sciences	Non-STEM degree
Total	41,530	9,262	1,359	646	3,706	3,551	32,268
STEM employment	4,736	3,327	763	167	1,738	659	1,409
Computer and math	2,167	1,331	637	120	447	128	835
Engineering	1,444	1,225	39	19	1,083	85	219
Physical and life sciences	654	484	8	9	54	413	170
STEM manager	471	287	80	19	155	33	184
Non-STEM employment	36, 794	5,935	595	479	1,968	2,892	30,859

Source: ESA calculations using American Community Survey public-use microdata.

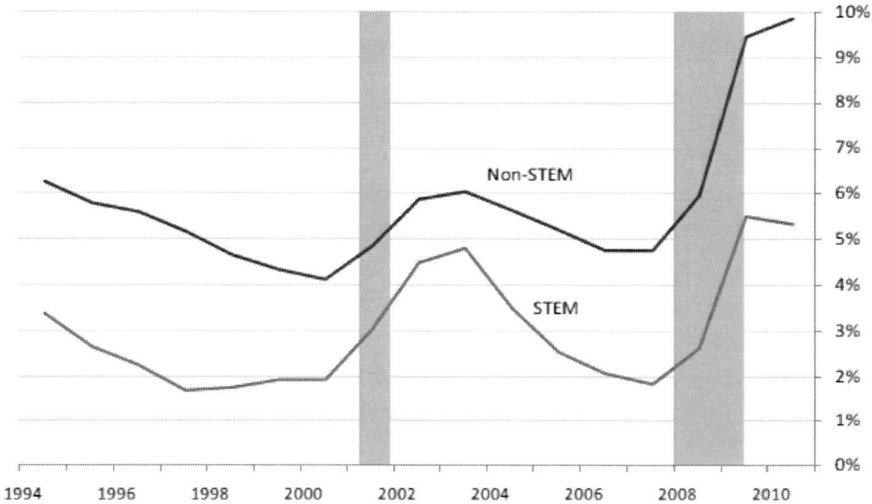

Source: ESA calculations using Current Population Survey public-use microdata.
Note: The estimates are for the civilian labor force age 16 and over. Shading indicates recession.

Figure 3. Unemployment Rates in STEM and Non-STEM Occupations, 1994-2010.

STEM JOBLESSNESS

In addition to higher earnings, workers in STEM occupations on average experience lower unemployment rates than workers in other fields (see Figure 3).[10] The unemployment rate for STEM workers rose from 1.8 percent in 2007 to 5.5 percent in 2009 before easing to 5.3 percent in 2010. The unemployment rate for non-STEM workers rose from 4.8 percent in 2007 to 9.5 percent in 2009 and then continued to increase to almost 10 percent in 2010. STEM workers, however, are not totally immune to economic downturns, as STEM joblessness did increase during the last two recessions.

Some of the difference in unemployment rates between STEM and non-STEM workers reflects differences in educational attainment. On balance, workers with a higher educational level tend to ex- perience lower unemployment, and STEM work- ers tend to be better educated. Looking at work- ers with a bachelor's degree or graduate degree, one finds less of a difference in unemployment rates between STEM and non-STEM workers than for those with less education. During the latest recession, the unemployment rate for col- lege-educated STEM workers edged above the non-STEM rate in 2009, but the rate for both groups converged to 4.7 percent in 2010 (see Figure 4). While college-educated STEM workers were less likely to be jobless than other workers during the latter part of the last two economic expansions, they were more likely to be jobless during and after the 2001 recession. The de- crease in the demand for information technol- ogy workers following the Y2K efforts and the crash of the Internet dot-com bubble likely played a role.

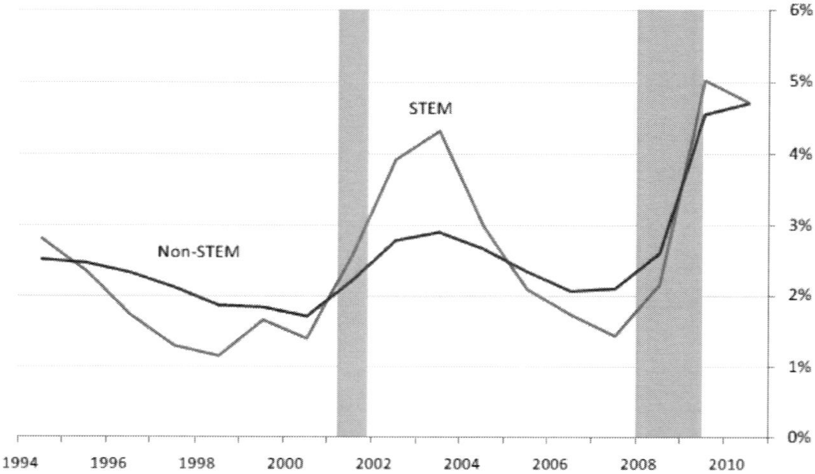

Source: ESA calculations using Current Population Survey public-use microdata.

Note: The estimates are for the civilian labor force age 25 and over with a bachelor's degree or higher. Shading indicates recession.

Figure 4. Unemployment Rates in STEM and Non-STEM Occupations, Workers with a Bachelor's Degree or Higher, 1994-2010.

Educational Attainment of STEM Workers

One of the more distinguishing characteristics of STEM workers is their educational attainment. More than two-thirds (68 percent) of STEM workers have a bachelor's degree or higher, compared to just under one-third (31 percent) of other workers age 16 and over (see Figure 5). Among the four STEM occupational groups, the physical and life sciences have the highest-educated workforce, with nearly 40 percent holding a graduate degree – about double the share for computer, math and engineering jobs. Nonetheless, because STEM includes professionals as well as first-tier support jobs, we find that a number of STEM workers have less than a four-year college degree; nearly one-quarter (23 percent) have completed an associate degree or at least some college, and 9 percent have a high school diploma or less. So while it is certainly true that the majority of STEM workers tend to have at least a bachelor's degree, opportunities also exist for STEM workers with lower education levels.

CONCLUSION

The greatest advancements in our society from medicine to mechanics have come from the minds of those interested in or studied in the areas of STEM. Although still relatively small in number, the STEM workforce has an outsized impact on a nation's competitiveness, economic growth, and overall standard of living. Analysis of data from the U.S. Census Bureau's American Community Survey and Current Population Survey provide new insights into the growing STEM workforce that is central to our economic vitality. STEM jobs are the jobs of the future. They are essential for developing our technological innovation and global competitiveness.

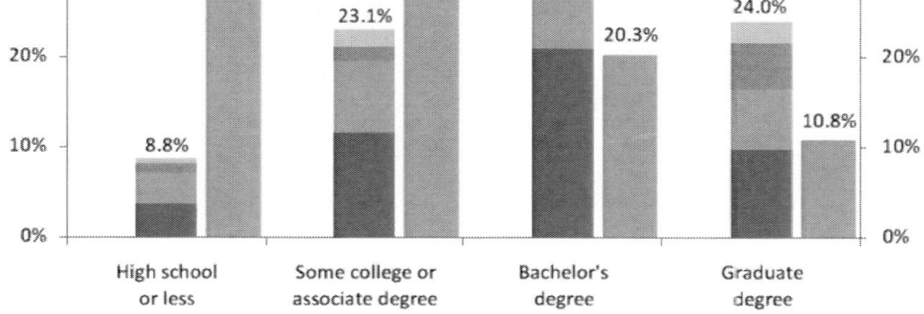

Source: ESA calculations using Current Population Survey public-use Source: ESA calculations using Current Population Survey public-use microdata.
Note: The estimates are for all employed persons age 16 and over.

Figure 5. Percent Distribution of STEM and Non-STEM Employment by Educational Attainment, 2010.

Appendix Table 1. Detailed STEM occupations and Standard Occupational Classification (SOC) codes

Occupation	SOC code	Occupation	SOC code
Computer and math occupations			
Computer scientists and systems analysts	15-10XX	Network systems and data communications analysts	15-1081
Computer programmers	15-1021	Mathematicians	15-2021
Computer software engineers	15-1030	Operations research analysts	15-2031
Computer support specialists	15-1041	Statisticians	15-2041
Database administrators	15-1061	Miscellaneous mathematical science occupa-tions	15-2090
Network and computer systems administra-tors	15-1071		
Engineering and surveying occupations			
Surveyors, cartographers, and photogrammet-rists	17-1020	Materials engineers	17-2131
Aerospace engineers	17-2011	Mechanical engineers	17-2141
Agricultural engineers	17-2021	Mining and geological engineers, including mining safety engineers	17-2151
Biomedical engineers	17-2031	Nuclear engineers	17-2161
Chemical engineers	17-2041	Petroleum engineers	17-2171
Civil engineers	17-2051	Engineers, all other	17-2199
Computer hardware engineers	17-2061	Drafters	17-3010
Electrical and electronic engineers	17-2070	Engineering technicians, except drafters	17-3020
Environmental engineers	17-2081	Surveying and mapping technicians	17-3031
Industrial engineers, including health and	17-2110	Sales engineers	41-9031

Occupation	SOC code	Occupation	SOC code
safety			
Marine engineers and naval architects	17-2121		
Physical and life sciences occupations			
Agricultural and food scientists	19-1010	Physical scientists, all other	19-2099
Biological scientists	19-1020	Agricultural and food science technicians	19-4011
Conservation scientists and foresters	19-1030	Biological technicians	19-4021
Medical scientists	19-1040	Chemical technicians	19-4031
Astronomers and physicists	19-2010	Geological and petroleum technicians	19-4041
Atmospheric and space scientists	19-2021	Nuclear technicians	19-4051
Chemists and materials scientists	19-2030	Other life, physical, and social science techni-cians	19-40XX
Environmental scientists and geoscientists	19-2040		
STEM managerial occupations			
Computer and information systems managers	11-3021	Natural sciences managers	11-9121
Engineering managers	11-9041		

These factors make STEM workers highly desirable to companies developing or operating on the technological forefront and extremely important to the U.S. economy, as competitive businesses are the foundation of a competitive economy. As this analysis highlights, STEM jobs should also be highly desirable to American workers. Regardless of educational attainment, entering a STEM profession is associated with higher earnings and reduced joblessness.

For college graduates, there is a payoff in choosing to pursue a STEM degree, and for America's workers, an even greater payoff in choosing a STEM career.

Appendix Table 2. Detailed STEM undergraduate majors

	Computer majors	
• Computer and information systems	• Computer science	• Computer administration management and security
• Computer programming and data processing	• Information sciences	• Computer networking and telecommunications
	Math majors	
• Mathematics	• Statistics and decision science	• Mathematics and computer science
• Applied mathematics		
	Engineering majors	
• General engineering	• Environmental engineering	• Petroleum engineering
• Aerospace engineering	• Geological and geophysical engineering	• Miscellaneous engineering
• Biological engineering	• Industrial and manufacturing engineering	• Engineering technologies
• Architectural engineering	• Materials engineering and materials science	• Engineering and industrial management
• Biomedical engineering	• Mechanical engineering	• Electrical engineering technology
• Chemical engineering	• Metallurgical engineering	• Industrial production technologies
• Civil engineering	• Mining and mineral engineering	• Mechanical engineering related technologies
• Computer engineering	• Naval architecture and marine engineering	• Miscellaneous engineering technologies
• Electrical engineering	• Nuclear engineering	• Military technologies
• Engineering mechanics physics and science		
	Physical and life sciences majors	
• Animal sciences	• Genetics	• Physical sciences
• Food science	• Microbiology	• Astronomy and astrophysics
• Plant science and agronomy	• Pharmacology	• Atmospheric sciences and meteorology
• Soil science	• Physiology	• Chemistry
• Environmental science	• Zoology	• Geology and earth science
• Biology	• Miscellaneous biology	• Geosciences
• Biochemical sciences	• Nutrition sciences	• Oceanography
• Botany	• Neuroscience	• Physics
• Molecular biology	• Cognitive science and biopsychology	• Nuclear, industrial radiology, and biological technologies
• Ecology		

End Notes

[1] These occupations are computer and information systems managers, engineering managers, and natural sciences managers.

[2] Although our principal data sources, the monthly Current Population Survey (CPS) and the 2009 American Community Survey (ACS), collect detailed information on workers'

occupations, they do not break out educators by their specific field. As a result, it is not possible to distinguish math and science professors from other professors. Data from the Bureau of Labor Statistics' Occupational Employment Statistics program show that there are roughly 200,000 postsecondary teachers in STEM fields, and so their exclusion is unlikely to materially affect our results.

3 The National Science Foundation does count social scientists among "science and engineering jobs" in keeping with the agency's mission supporting "all fields of fundamental science and engineering, except for medical sciences."

4 The distinction between "scientists" rather than "science occupations" is more than just semantic as "science occupations" covers not just scientists but also science technicians. Likewise, engineering and surveying occupations include engineering technicians and drafters, and computer occupations range from computer support specialists to computer software engineers.

5 Using 2008-18 employment projections from the Bureau of Labor Statistics (BLS), ESA calculated the projected employment growth of STEM occupations. BLS's Employment Projections Program's homepage is http:// www.bls.gov/emp and detailed occupational projections are available at http://www.bls.gov/emp/ eptable106.htm

6 More specifically, the earnings regressions control for age (up to a fourth degree polynomial of age), gender, marital status, race and Hispanic origin, nativity and citizenship, educational attainment, metropolitan area, region, union representation, major industry, STEM occupation, time, and STEM occupation interacted with time.

7 For the regression analysis, we combined workers that had completed some college, high school, or less into a single "less than a bachelor's degree" category because of small sample concerns.

8 One caveat with these results is that the STEM premium may also capture other unidentified factors that systematically distinguish STEM workers from other workers. STEM workers may have more in common than just their career fields, such as a similar work ethic or affinity, and the STEM premium could reflect the net impact of those unidentified common characteristics as well as the pure premium from working in a STEM job.

9 As mentioned earlier, a person whose primary or secondary major was in a STEM field is counted as having a STEM undergraduate degree.

10 In the Current Population Survey, occupations are assigned to persons based on their most recent work experience. As a result, unemployment rates by occupation are sometimes referred to as the "experienced unemployment rate." Thus, if we define USTEM as the number of unemployed persons whose most recent job was in a STEM occupation and ESTEM as the number of persons currently employed in a STEM occupation, then the STEM unemployment rate is calculated as USTEM / (ESTEM + USTEM).

In: STEM Workforce Trends … ISBN: 978-1-62257-818-4
Editors: C. Martin Garcia and S. Cook

Chapter 3

WOMEN IN STEM: A GENDER GAP TO INNOVATION*

United States Department of Commerce

EXECUTIVE SUMMARY

Our science, technology, engineering and math (STEM) workforce is crucial to America's innovative capacity and global competitiveness. Yet women are vastly underrepresented in STEM jobs and among STEM degree holders despite making up nearly half of the U.S. workforce and half of the college-educated workforce. That leaves an untapped opportunity to expand STEM employment in the United States, even as there is wide agreement that the nation must do more to improve its competitiveness.

- Although women fill close to half of all jobs in the U.S. economy, they hold less than 25 percent of STEM jobs. This has been the case throughout the past decade, even as college-educated women have increased their share of the overall workforce.
- Women with STEM jobs earned 33 percent more than comparable women in non-STEM jobs – considerably higher than the STEM

* This is an edited, reformatted and augmented version of the U.S. Department of Commerce, Economics and Statistics Administration, dated August 2011.

premium for men. As a result, the gender wage gap is smaller in STEM jobs than in non-STEM jobs.

- Women hold a disproportionately low share of STEM undergraduate degrees, particularly in engineering.
- Women with a STEM degree are less likely than their male counterparts to work in a STEM occupation; they are more likely to work in education or healthcare.
- There are many possible factors contributing to the discrepancy of women and men in STEM jobs, including: a lack of female role models, gender stereotyping, and less family-friendly flexibility in the STEM fields. Regardless of the causes, the findings of this report provide evidence of a need to encourage and support women in STEM.

WOMEN IN STEM JOBS

According to the Census Bureau's 2009 American Community Survey (ACS), women comprise 48 percent of the U.S. workforce but just 24 percent of STEM workers. In other words, half as many women are working in STEM jobs as one might expect if gender representation in STEM professions mirrored the overall workforce. (See Figure 1.)

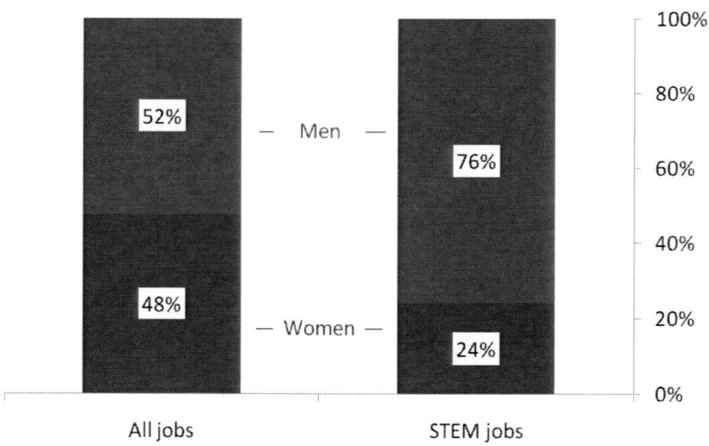

Source: ESA calculations from American Community Survey public-use microdata.
Note: Estimates are for employed persons age 16 and over.

Figure 1. Gender Shares of Total and STEM Jobs, 2009.

This underrepresentation has remained fairly constant over the past decade, even as women's share of the college-educated workforce has increased. As shown in Table 1, between 2000 and 2009, women's share of the STEM workforce remained constant at 24 percent, while their share of all college-educated workers increased from 46 to 49 percent. Using data from the Census Bureau's Current Population Survey (CPS) going back to 1994, almost identical trends emerge.

Among STEM jobs, women's representation has varied over time. While the female share has declined in computer and math jobs, their share has risen in other occupations. In 2009, women comprised 27 percent of the computer and math work-force (the largest of the four STEM components), a drop of 3 percentage points since 2000. (See Table 2.) Engineers are the second largest STEM occupational group, but only about one out of every seven engineers is female. Interestingly, the number of female engineers edged up by 12,000 over nine years, while the number of male engineers declined by 106,000. In physical and life sciences jobs, however, women made up about 40 percent of the workforce in 2009, up from 36 percent in 2000. STEM managers is another area that has shown growth, with women's share of the workforce increasing to 25 percent.

Men are much more likely than women to have a STEM job regardless of educational attainment. Figure 2 demonstrates that higher education levels generally correspond to an increased likelihood of having a STEM job for both men and women. The gap lessens somewhat at the doctoral level, but women still lag far behind men in STEM employment.[1]

Table 1. Total and STEM Employment by Gender and Educational Attainment, 2000 and 2009

(thousands of workers)	Male		Female		Percent Female	
	2000	**2009**	**2000**	**2009**	**2000**	**2009**
All workers	69,098	73,580	60,619	67,058	47%	48%
College-educated	18,995	22,167	16,415	21,433	46%	49%
STEM workers	5,321	5,640	1,680	1,790	24%	24%
College-educated	3,259	3,738	1,002	1,199	24%	24%

Source: ESA calculations from Census 2000 and 2009 American Community Survey public-use microdata.

Note: Estimates are for employed persons age 16 and over.

Table 2. Employment in STEM Occupations in 2009

(thousands of workers)						
	Male		**Female**		**Percent Female**	
	2000	**2009**	**2000**	**2009**	**2000**	**2009**
STEM total	5,321	5,640	1,680	1,790	24%	24%
Computer science and math	2,202	2,534	940	929	30%	27%
Engineering	2,185	2,079	318	330	13%	14%
Physical and life sciences	551	553	310	374	36%	40%
STEM managers	382	474	111	157	23%	25%

Source: ESA calculations from Census 2000 and 2009 American Community Survey public-use microdata.

Note: Estimates are for employed persons age 16 and over.

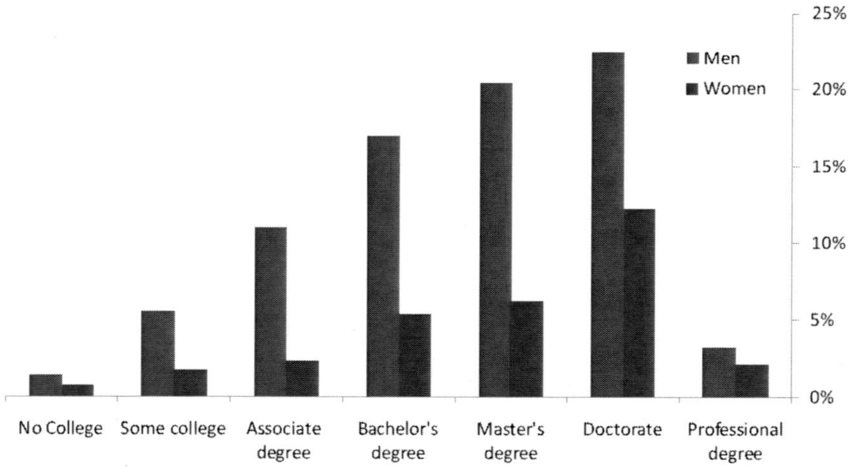

Source: ESA calculations from American Community Survey public-use microdata.

Note: Professional degrees include medical, dental, veterinary, and law degrees. Estimates are for employed persons age 16 and over.

Figure 2. Share of Workers in STEM Jobs by Gender and Educational Attainment, 2009.

STEM Worker Earnings and Gender

There are two notable findings in examining the relationship between STEM, gender and earnings. First, as is clearly documented in our previous report "*STEM: Good Jobs Now and for the Future,*" STEM workers earn considerably more than their non-STEM counterparts (what we call in this report the "STEM earnings premium"). Second is the gender wage gap – a robust finding that women earn considerably less than men, even after controlling for a wide set of characteristics such as education and age. Figure 3 highlights how these two findings intersect by showing the average hourly earnings of full-time, year-round workers in private sector STEM and non-STEM jobs. On average, men and women earn $36.34 and $31.11 per hour, respectively, in STEM jobs – higher than the $24.47 that men earn and $19.26 that women earn, on average, in other occupations. For every dollar earned by a man in STEM, a woman earns 14 cents (or 14 percent) less, smaller than the 21 percent gender wage gap in non-STEM occupations, but a clear gender disparity nonetheless.

While illustrative, these simple comparisons do not take into account the many factors that can help explain why STEM workers tend to earn more than non-STEM workers, or why women earn less than men. Following the methodology of our earlier STEM report, we use regression analyses to control for many of these factors, including workers' age, educational attainment, and region of residence. The results of these analyses underscore both the STEM earnings premium and the gender earnings gap.

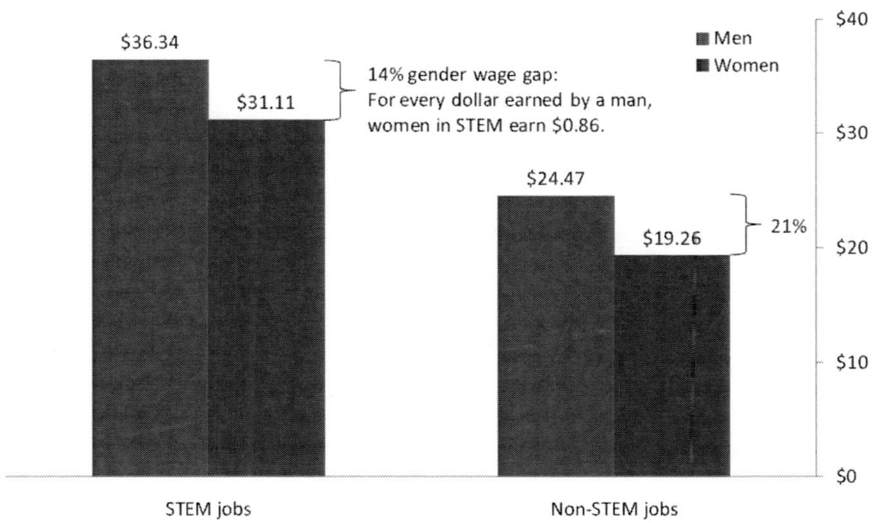

Source: ESA calculations from American Community Survey public-use microdata.

Note: Estimates are for full-time year-round private wage and salary workers age 16 and over.

Figure 3. Average Hourly Earnings by Gender and Occupation, 2009.

In our previous report on this issue, we found that STEM workers earn significantly more than their non-STEM counterparts in the private sector. So, one way to compare men and women in STEM is to see to what extent their STEM earnings premium varies. Our analysis shows that, all else being equal, women in STEM jobs earn 33 percent more than their female peers in other jobs, while the STEM premium for men is 25 percent.

Given that the STEM premium for women is higher than for men, we would expect women in STEM to face a smaller gender earnings gap than women in other occupations. Not surprisingly, we find that the gap shrinks but does not disappear for women in STEM. Furthermore, as we look at more specific groups of STEM workers (which allows us to make better "apples-to-apples" comparisons of male and female wages), we find an even smaller gender wage difference. Figure 4 shows the regression-adjusted gender wage gap for college-educated STEM workers in each of the four major occupational groups. Interestingly, the most male-dominated STEM occupational group—engineers—is also the one with the smallest regression-adjusted wage gap; female engineers earned 7 percent less per hour than male counterparts. Physical and life sciences occupations, the most gender-balanced STEM group, have an 8 percent wage gap, and STEM managers a 9 percent gap. Notably higher was the 12 percent gender wage gap in computer and math occupations.

STEM Degrees and Fields of Study by Gender

Since the gateway to many high-paying STEM jobs is a STEM degree, it is useful to examine the extent to which college-educated workers had STEM degrees. The 2009 ACS provides a rich new data source for analyzing the link between undergraduate studies and subsequent employment. The ACS data on undergraduate fields of study show that women account for nearly half of employed college graduates age 25 and over, but only about 25 percent of employed STEM degree holders and an even smaller share – just about 20 percent – of STEM degree holders working in STEM jobs.

There were 2.5 million college-educated working women with STEM degrees in 2009 compared with 6.7 million men. What makes this disparity even more alarming is that, in the overall labor force, there are 21.4 million women (49 percent of the total) and 22.2 million men who are employed and have bachelor's degrees.

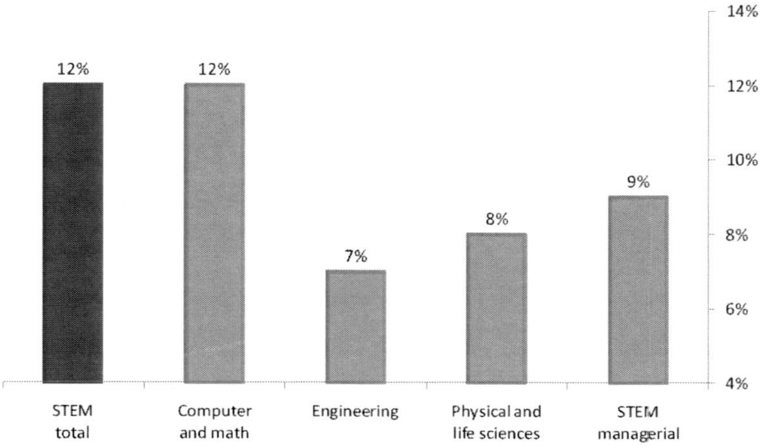

Source: ESA calculations from American Community Survey public-use microdata.

Note: Estimates are for full-time year-round private wage and salary workers age 25 and over.

Figure 4. Regression-adjusted Gender Wage Gap of College-educated STEM Workers by Occupation, 2009.

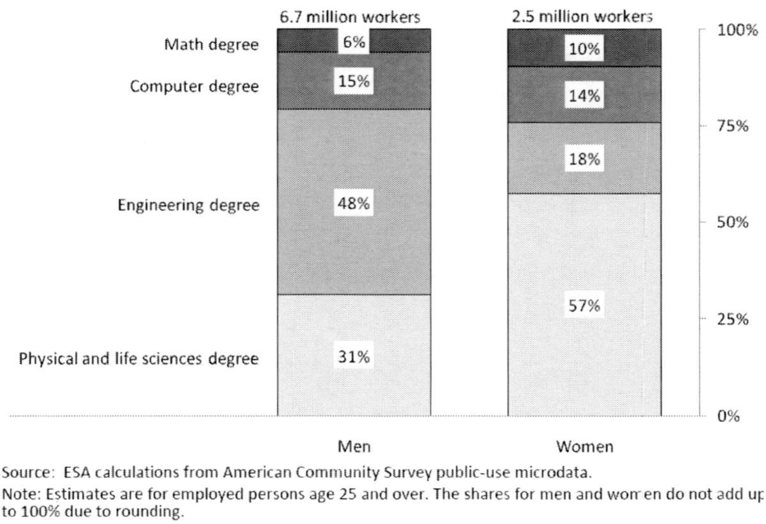

Source: ESA calculations from American Community Survey public-use microdata.

Note: Estimates are for employed persons age 25 and over. The shares for men and women do not add up to 100% due to rounding.

Figure 5. College-educated Workers with a STEM Degree by Gender and STEM Degree Field, 2009.

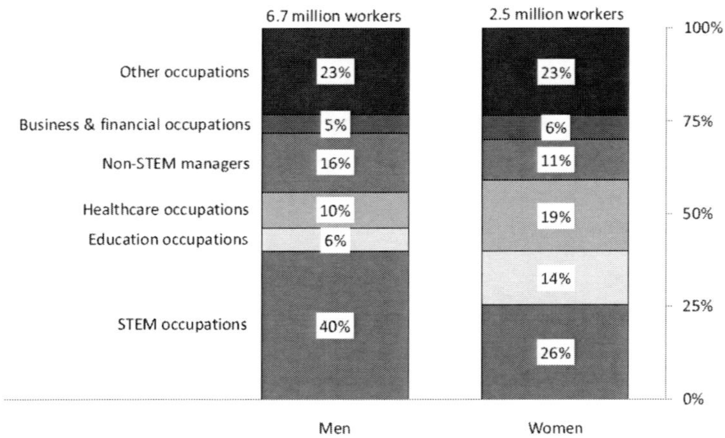

Source: ESA calculations from American Community Survey public-use microdata.
Note: Estimates are for employed persons age 25 and over. The shares for men and women do not add up to 100% due to rounding.

Figure 6. College-educated Workers with a STEM Degree by Gender and STEM Occupation, 2009.

Among STEM majors, the distribution of men and women differs significantly. As shown in Figure 5, well over half (57 percent) of female STEM majors study physical and life sciences, while fewer than one-third (31 percent) of men choose these fields. The share of women choosing math majors is also higher than men: 10 versus 6 percent. The bulk of men with STEM majors (48 percent) choose engineering degrees, two-and-a-half times the share of women who choose engineering. Equal shares of male and female STEM majors enter computer science. In terms of raw numbers, however, men in the workforce with STEM degrees outnumber women across all four fields of study.[2]

STEM DEGREES AND CAREERS BY GENDER

As noted in the previous section, college-educated women are much less likely than men to major in STEM fields. But even when women choose STEM degrees, their typical career paths diverge substantially from their male counterparts. About 40 percent (2.7 million) of men with STEM college degrees work in STEM jobs, whereas only 26 percent (0.6 million) of women with STEM degrees work in STEM jobs. (See Figure 6.)

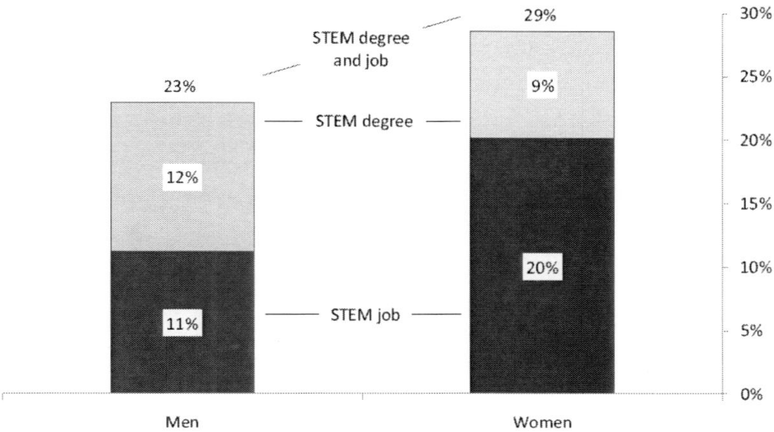

Source: ESA calculations from American Community Survey public-use microdata.

Note: Estimates are for full-time year-round private wage and salary workers who have at least a bachelor's degree and are age 25 and over.

Figure 7. Wage Premium from Having a STEM Job and/or Degree.

Men are more likely to have non-STEM management jobs than women, 16 percent and 11 percent, respectively. In contrast, female STEM majors are twice as likely as men to work in education or healthcare. Nearly one in five STEM college-educated women works in healthcare occupations, compared with about one in ten men. Likewise, approximately 14 percent of female STEM majors end up in education occupations, compared with approximately 6 percent of men. Similar shares of men and women with STEM degrees worked in business and financial occupations or other fields.

STEM Premiums and Gender

As highlighted in our earlier STEM report, receiving a STEM degree tends to result in higher earnings later in life. Figure 7 illustrates the considerable extent to which the earnings premium from having a STEM job or STEM degree varies by gender. Women enjoy a much bigger STEM job premium than men, but a slightly smaller premium for having earned a STEM bachelor's degree. Specifically, when we control for whether or not women have STEM degrees, we find that college-educated women (regardless of choice of undergraduate major) earn 20 percent more in STEM jobs than elsewhere. This is nearly double the 11 percent premium that college-educated men realize working in STEM. On the other hand, female STEM degree holders earn 9 percent more than women with other degrees, regardless of their job. The STEM degree premium for men is somewhat higher at nearly 12 percent. The biggest STEM-related wage premiums go to men and women who both major in a STEM field and choose a STEM job. This career path nets women 29 percent higher hourly earnings, on average, than their peers who have neither a STEM degree nor a STEM job. The corresponding premium for men is smaller, but also sizeable, at 23 percent.

CONCLUSION

This report finds that women are underrepresented both in STEM jobs and STEM undergraduate degrees and have been consistently over the last decade. The relatively few women who receive STEM degrees are concentrated in physical and life sciences, in contrast to men, who are concentrated primarily in engineering. Women who do receive STEM degrees are less likely to work in STEM jobs than their male counterparts. And while women working in STEM jobs earn less than their male counterparts, they experience a smaller gender wage gap compared to others in non-STEM occupations.

The underrepresentation of women in STEM majors and jobs may be attributable to a variety of factors. These may include different choices men and women typically make in response to incentives in STEM education and STEM employment – for example, STEM career paths may be less accommodating to people cycling in and out of the work-force to raise a family – or it may be because there are relatively few female STEM role models. Perhaps strong gender stereotypes discourage women from pursuing STEM education and STEM jobs.

While this report does not – and cannot – explain why gender differences in STEM exist, it does aim to provide data and insight that will enable more informed policymaking. The findings provide definitive evidence of a need to encourage and support women in STEM with a goal of gender parity. Given the high-quality, well-paying jobs in the fields of science, technology, engineering and math, there is great opportunity for growth in STEM in support of American competitiveness, innovation and jobs of the future.

End Notes

[1] Few persons with professional degrees (e.g., MBA's, law, and medicine) work in STEM jobs because such persons typically have jobs that we do not characterize as STEM.

[2] Using data from a different source, the number of bachelor's degrees conferred annually in STEM fields in the United States increased about 20 percent between 1998 and 2008 (from 200,000 to 240,000). But women's share has not risen and remains far below parity with men (35 percent in both years), despite the fact that a majority of all bachelor degrees are conferred on women (56 percent in 1998 and 57 percent in 2008). About one-quarter of bachelor's degrees awarded to men are in STEM fields, compared with one-tenth conferred on women; these shares have remained about the same between 1998 and 2008. About two-thirds of STEM degrees awarded to men were in engineering and computer science, compared with one-quarter awarded to women. Over half of STEM degrees awarded to women were in biological sciences compared with 20 percent for men. See U.S. Department of Commerce, Economics and Statistics Administration and Executive Office of the President, Office of Management and Budget (2011) *Women in America: Indicators of Social and Economic Well-Being* (http://www.esa.doc.gov/sites/default/files/reports/documents/womeninamerica.pdf), p. 23.

In: STEM Workforce Trends ... ISBN: 978-1-62257-818-4
Editors: C. Martin Garcia and S. Cook

Chapter 4

EDUCATION SUPPORTS RACIAL AND ETHNIC EQUALITY IN STEM*

United States Department of Commerce

EXECUTIVE SUMMARY

Science, technology, engineering, and math (STEM) workers are essential to American innovation and competitiveness in an increasingly dynamic and global marketplace. In this third report, we examine demographic disparities in STEM education and find that educational attainment may affect equality of opportunity in these critical, high-quality jobs of the future.

This report follows an analysis of labor market outcomes and gender disparities among STEM workers. We find that regardless of race and Hispanic origin, higher college graduation rates are associated with higher shares of workers with STEM jobs. But non-Hispanic Whites and Asians are much more likely than other minority groups to have a bachelor's degree. By increasing the numbers of STEM workers among currently underrepresented groups through education we can help ensure America's future as a global leader in technology and innovation.

Other key findings of this report include:

* This is an edited, reformatted and augmented version of the U.S. Department of Commerce Economics and Statistics Administration, dated September 2011.

- Non-Hispanic Whites comprise the largest group of STEM workers, accounting for about seven out of ten STEM workers, which aligns closely with their share of the overall workforce.
- Non-Hispanic Asians are most likely (42 percent) to graduate college with a STEM degree, while the propensities of other groups are all fairly similar (17-22percent).
- Half of all non-Hispanic Asian workers with STEM degrees have STEM jobs, compared to 30 percent of Hispanics and non-Hispanic Black and American Indian and Alaska Native workers.
- One in five STEM workers is foreign-born, of which 63 percent come from Asia.
- STEM workers in all demographic groups, including the foreign-born, earn more than their non-STEM counterparts. Hispanics and non-Hispanic Blacks receive a significantly larger STEM premium than do non-Hispanic Whites.

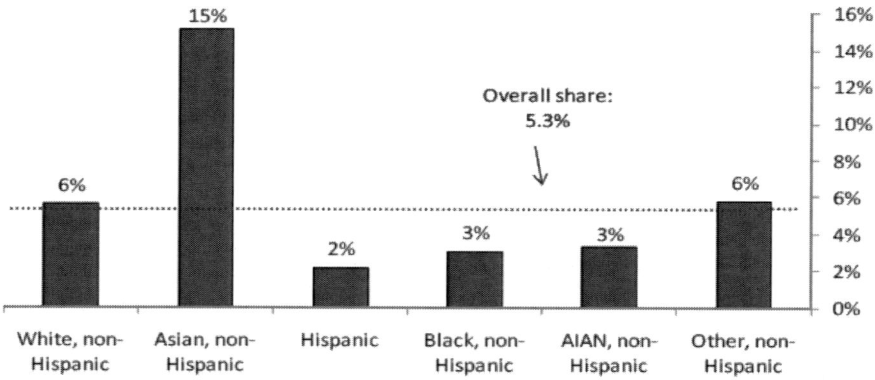

Source: ESA calculations from American Community Survey public-use microdata.

Note: Estimates are for employed persons age 16 and over. See box on next page for definitions of race and Hispanic origin used in this report.

Figure 1. Share of Workers with STEM Jobs, by Race and Hispanic Origin, 2009.

The Commerce Department's Economics and Statistics Administration has released two other reports on STEM workers. The first, *STEM: Good Jobs Now and for the Future*, was released in July 2011. The second, *Women in STEM: A Gender Gap to Innovation*, was released in August 2011.

This report uses six non-overlapping groups to study STEM workers by race and Hispanic origin: White alone, not Hispanic; Black alone, not Hispanic; Asian alone, not Hispanic; American Indian or Alaska Native alone, not Hispanic (AIAN); Other, not Hispanic; and Hispanic of any race. The "Other" race category includes Native Hawaiian or Other Pacific Islanders; other races not listed in this report; and people who two or more races. In some figures, AIAN Other are combined in a category called "All other" when the number of AIAN observations is too small.

STEM Employment by Race and Hispanic Origin

According to the U.S. Census Bureau's 2009 American Community Survey (ACS), almost three out of four STEM jobs (72 percent) are held by non-Hispanic Whites (See Figure 2), which is close to their overall representation in the U.S. workforce (68 percent). In contrast, non-Hispanic Asians make up 14 percent of all STEM workers but only five percent of the U.S. workforce — almost three times as many non-Hispanic Asians hold a STEM job as would be expected based on their overall representation in the workforce. However, only half as many non-Hispanic Black or Hispanic workers have STEM jobs relative to their overall representation in the U.S. workforce. (Non-Hispanic Blacks and Hispanics each account for six percent of all STEM workers, but 11 percent and 14 percent, respectively, of overall employment.)

Over the past decade, the share of non-Hispanic Whites in the STEM workforce has fallen about six percentage points (from 78 percent to 72 percent as shown in Figure 2), as did their share of the overall workforce. Hispanics have accounted for most of the increase in the overall U.S. workforce since 2000 (their share rose from 10 percent in 2000 to 14 percent in 2009), yet the Hispanic share of the STEM workforce increased by only one percentage point (from five percent to six percent). Conversely, the non-Hispanic Asian share of STEM workers increased by four percentage points (from 10 percent to 14 percent), whereas their share of the overall workforce rose by only one percentage point (from 10 percent to 11 percent). Non-Hispanic Blacks did not see any change in their share of STEM workers, while their share of the overall workforce increased by one percentage point (from 10 percent to 11 percent).

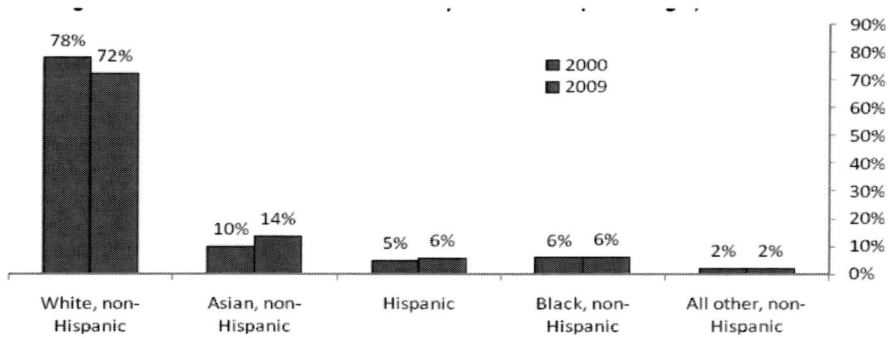

Source: ESA calculations from 2000 Census and American Community Survey public-use microdata.

Note: Estimates are for employed persons age 16 and over. The data for each year may not sum to 100% due to rounding.

Figure 2. Distribution of STEM Workers by Race and Hispanic Origin, 2000 and 2009.

STEM Employment and Education by Race and Hispanic Origin

The propensity for workers to have STEM jobs is driven by three key decisions made along their career paths. First, Hispanics, non-Hispanic Blacks and American Indians or Alaska Natives (AIANs) are much less likely to obtain a bachelor's or higher degree in comparison with non-Hispanic Whites and Asians. Second, Hispanics and non-Hispanic Whites and Blacks who obtain college degrees are less likely than non-Hispanic Asians to major in STEM fields (although as we found in our earlier STEM reports, many college-educated STEM workers do not have STEM degrees). And third, STEM majors among Hispanics and non-Hispanic Blacks, Whites, and members of all other groups are less likely to obtain STEM jobs than non-Hispanic Asians.

The propensity to obtain a college degree (or higher) varies considerably by race and Hispanic origin. (See leftmost group of bars in Figure 3.) Only about one-fifth of non-Hispanic Blacks (22 percent) and less than one-sixth of Hispanics (14 percent) have bachelor's degrees. More than half (54 percent) of Asians and more than one-third(35percent) of non-Hispanic Whites, however, have at least a bachelor's degree. (Only 17 percent of AIANs have bachelor's degrees.)

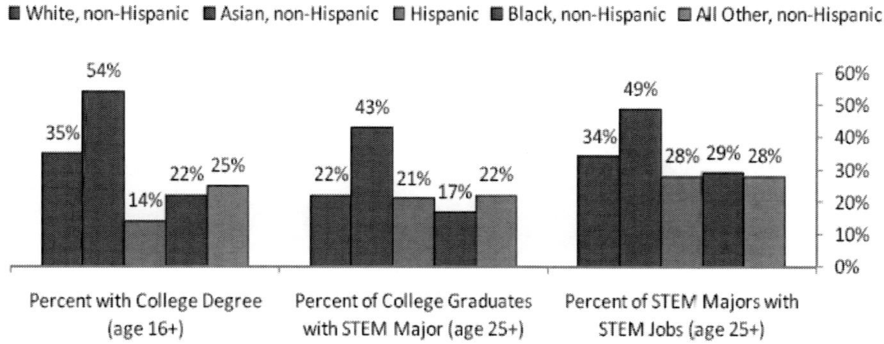

Source: ESA calculations from American Community Survey public-use microdata.
Note: Estimates are for employed persons.

Figure 3. Factors Affecting Likelihood of Having STEM Job by Race and Hispanic Origin, 2009.

The likelihood of having a STEM job is significantly higher among college-educated workers than those without a college degree, regardless of race and Hispanic origin. (See Figure 4.) One out of four non-Hispanic Asians with a bachelor's degree or higher has a STEM job, but only 3 percent of those without such a degree have a STEM job. Ten percent of non-Hispanic Whites with a bachelor's degree or higher have STEM jobs – only two percentage points higher than is the case for Hispanics and non-Hispanic Blacks and AIANs.

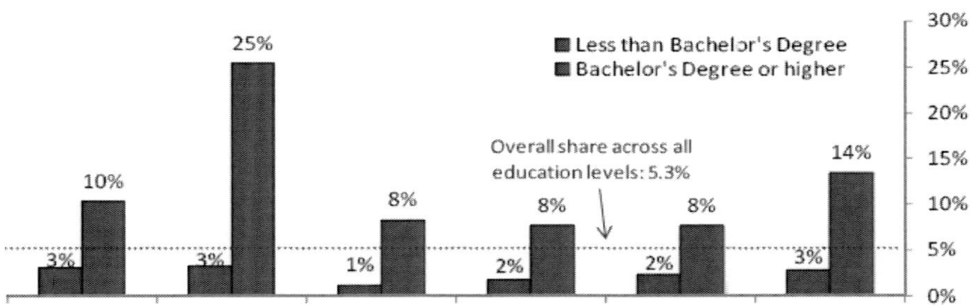

Source: ESA calculations from American Community Survey public-use microdata.
Note: Estimates are for employed persons age 16 and over. *AIAN refers to persons who identify as American Indian or Alaskan Native alone, not Hispanic.

Figure 4. Share of Workers with STEM Jobs by Race, Hispanic Origin, and Education, 2009.

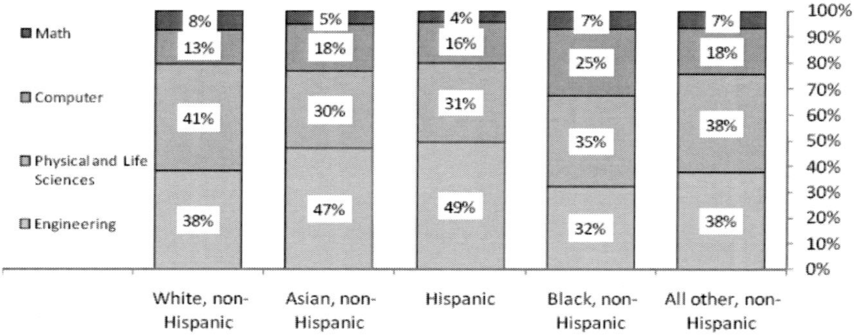

Source: ESA calculations from American Community Survey public-use microdata.

Figure 5. Degree Fields of Employed STEM Majors, by Race and Hispanic Origin, 2009.

The propensity for college-educated workers to pick a STEM major varies across race and Hispanic origin, though. (See middle group of bars in Figure 3.) Out of the 41.5 million workers with at least a bachelor's degree, 9.3 million (or 22 percent) have a STEM degree. Among workers with a bachelor's degree or higher, Hispanics and non-Hispanic Whites and members of all other groups are similarly represented in STEM majors (ranging between 21 to 22 percent), while non-Hispanic Blacks have a lower likelihood (17 percent) of having a STEM degree. Non-Hispanic Asians have the highest likelihood of having a STEM degree (43 3percent).

Engineering, and physical and life sciences are the most popular fields of study for college-educated workers with a STEM degree, together accounting for 67-to-80 percent of that workforce. (See Figure 5.) Among STEM degree categories, nearly half of all college-educated non-Hispanic Asians and Hispanics with a STEM degree majored in engineering, compared to 38 percent for non-Hispanic Whites and 32 percent for non-Hispanic Blacks. The share of STEM majors with a degree in physical or life sciences ranged between 30 and 41 percent of STEM majors across race and Hispanic origin. Also noteworthy is that non-Hispanic Blacks had a higher propensity to major in computer science than did the other groups.

Looking at those with STEM degrees, fewer than three out of ten Hispanics, non-Hispanic Blacks and members of all other groups have STEM jobs (28-29 percent) compared with one-third (34 percent)of non-Hispanic Whites and nearly half (49 percent) of non-Hispanic Asians (see rightmost group of bars in Figure 3). However, about the same shares of workers with STEM degrees in each demographic group had jobs in financial, healthcare, and education occupations. Whites with STEM degrees are somewhat more likely to be in non-STEM management positions relative to other groups.

Data indicate that one of the most important factors explaining the relatively low propensity of non-Hispanic Blacks and Hispanics to have STEM jobs is their relatively low college graduation rates.

FOREIGN-BORN STEM WORKERS

According to the 2009 ACS, there are about 1.5 million STEM workers in the U.S. who are foreign born – about 20 percent of the STEM workforce, which is slightly higher than the 16 percent share of foreign-born among all U.S. workers. (See Figure 6.) About half of foreign-born STEM workers are naturalized citizens (53 percent) and half noncitizens (47 percent). The share of foreign-born, however, varies significantly by race and Hispanic origin. Non-Hispanic Whites, the largest group of workers —representing almost three-fourths of all STEM workers as shown in Figure 2 — are mostly native born (93 percent). Non-Hispanic Asian STEM workers, in contrast, are almost all foreign-born. Nearly nine out of ten non-Hispanic Asian STEM workers (87 percent) were born outside of the United States. This trend also applies to all non-Hispanic Asian workers, the vast majority of whom (81 percent) are foreign-born. One out of six non-Hispanic Black STEM workers is foreign-born.

Unlike the non-Hispanic groups, foreign-born workers represent a smaller share of Hispanic STEM workers (36 percent) than of all Hispanic workers (54 percent). This largely reflects the fact that while Mexico accounts for more than 60 percent of foreign-born Hispanic workers, less than one percent of the foreign-born from Mexico work in STEM occupations. This low representation in STEM employment is also evident among foreign-born Hispanic workers from most of Central America as well as the Dominican Republic. However, Hispanic workers from other countries tend to be more broadly represented in STEM occupations. In particular, Hispanics born in some South American countries such as Venezuela (8.6 percent) and Chile (8.0 percent), and those from Spain (8.3 percent) are considerably more likely to work in STEM jobs than all Hispanics (2.2 percent) or foreign-born Hispanics (1.4 percent).

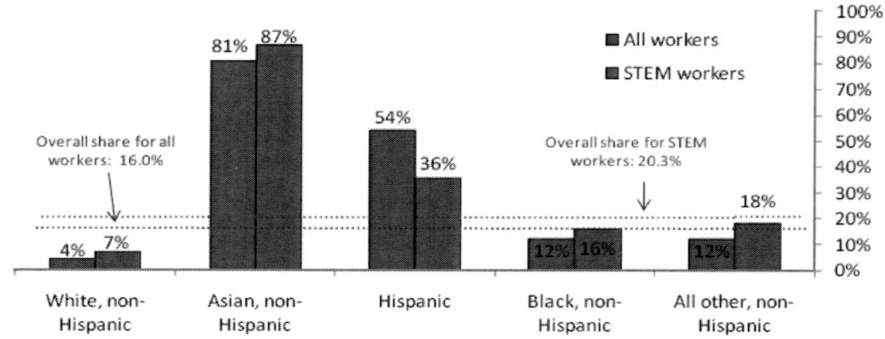

Source: ESA calculations from American Community Survey public-use microdata.

Figure 6. Foreign-born Share of All and STEM Workers, by Race and Hispanic Origin, 2009.

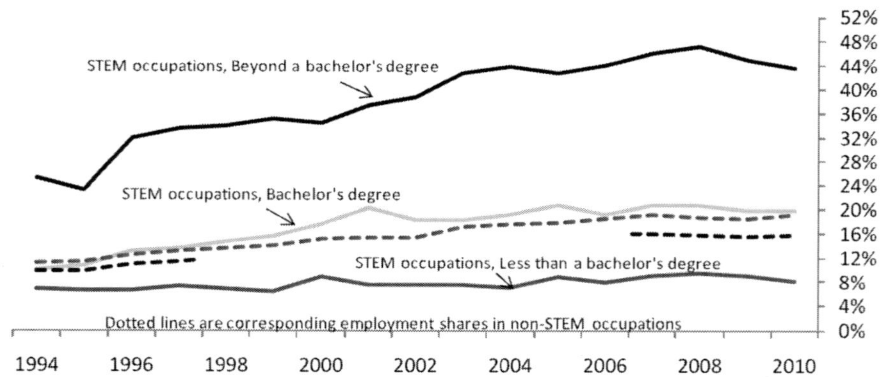

Source: ESA calculations using Current Population Survey public-use microdata files for 1994 through 2010.

Note: Estimates are for private full-time wage and salary workers age 25 and over.

Figure 7. Foreign-born Share of STEM Workers, by Educational Attainment, 1994-2010.

The foreign-born share of STEM workers with graduate degrees (44 percent) is about double the foreign-born share of STEM workers for all education levels. Moreover, the foreign-born share of STEM workers with graduate degrees has nearly doubled over the past 17 years, as has the foreign-born share of STEM workers with just a bachelor's degree (see Figure 7), which may in part be due to increasing global competition in STEM. Among STEM workers who do not have college degrees (only 8 percent), foreign-born workers are under-represented. Some of the increase in the foreign-born share of highly educated STEM workers may be due to a combination of the relatively high quality of U.S. colleges and universities – one of America's competitive advantages - and rising incomes in other countries that enable foreign students to pay for a high-quality U.S. STEM education. Quantifying these effects, however, is beyond the scope of this report.

More than half of all foreign-born workers in the United States were born in Latin America (55 percent), while only 27 percent of all foreign-born workers are from Asia. (See Figure 8.) However, these numbers are nearly reversed among foreign-born STEM workers, almost two-thirds (63 percent) of whom come from Asia. One in four foreign-born STEM workers come from China, while 11 percent come from India; no other country accounts for more than 5 percent of foreign-born STEM workers.

STEM Worker Earnings by Race and Hispanic Origin

Non-Hispanic White STEM workers on average earn 22 percent more than their non-STEM counter-parts, based on regression analysis that controls for other characteristics that may be related to earnings.[1] The STEM earnings premium is larger for other demographic groups -31 percent for non-Hispanic Asians, 39 percent for non-Hispanic Blacks, and 36 percent for Hispanics. (See Figure 9.) This indicates that non-Hispanic White workers, representing the majority (72 percent) of STEM workers in the U.S., get a significantly smaller boost from going into a STEM occupation than their otherwise similar Hispanic or non-Hispanic Asian and Black counterparts.

The STEM earnings premium is 24 percent for native-born STEM workers and 35 percent for foreign-born STEM workers. The boost in earnings associated with entering a STEM profession is about one-third larger among foreign-born non-citizens (41 percent) than foreign-born naturalized citizens (30 percent).

A portion of the wage premium from having a STEM job may be explained by occupational choices that vary across or within demographic groups that are not captured by the aggregate data on major industries. Some of the wage premium may also be due to other unidentified attributes that are simply not quantifiable. As a result, the regression-adjusted estimates for STEM wage premiums may vary depending on the exact specification of the regression model, that is, how much information we have on worker characteristics that likely impact earnings.

Our earlier STEM report, *"STEM: Good Jobs Now and for the Future,"* documented that in addition to earning a wage premium, STEM workers are also generally less likely to be unemployed than non-STEM workers, although the difference in the unemployment rate is much less among college-educated STEM and non-STEM workers. We find a similar pattern by race and Hispanic origin.

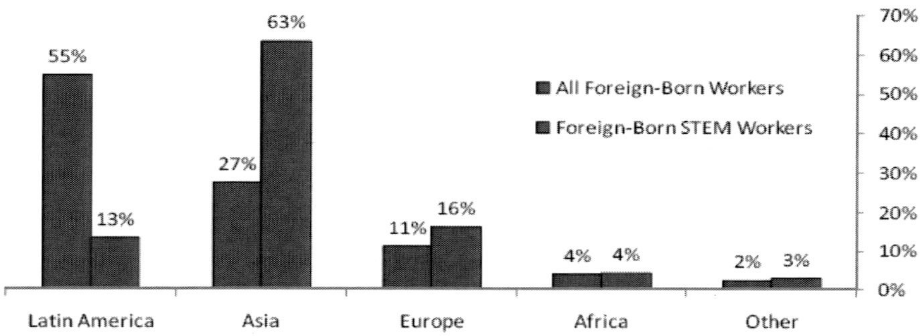

Source: ESA calculations from American Community Survey public-use microdata.
Note: Estimates are for employed persons age 16 and over.

Figure 8. All Foreign-born Workers and STEM Foreign-Born Workers, by Birthplace, 2009.

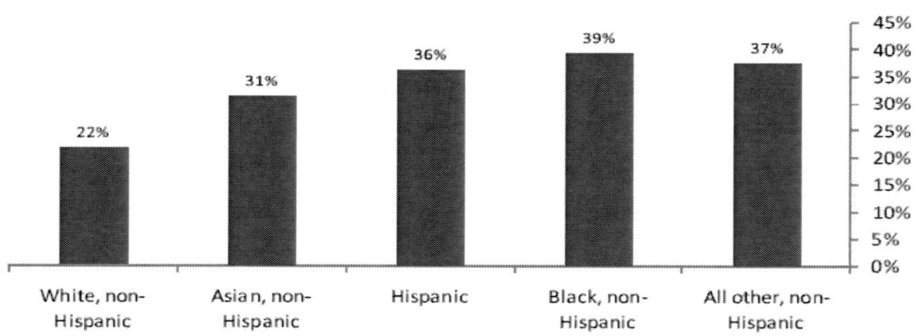

Source: ESA calculations from American Community Survey public-use microdata.
Note: Estimates are for full-time year-round private wage and salary workers age 25 and over.

Figure 9. Regression-adjusted Wage Premium from Having STEM Job, by Race and Hispanic Origin, 2009.

CONCLUSION

This report finds that non-Hispanic Blacks and Hispanics have been consistently underrepresented in STEM jobs over the past decade. Non-Hispanic Black and Hispanic workers are half as likely as all workers to have STEM jobs. In contrast, non-Hispanic Asian workers are nearly three times as likely as all workers to have STEM jobs. The variations observed across demographic groups may be attributable to a variety of factors, including differences among college graduation rates, propensities to choose a STEM field of study and preferences and other unobservable factors.

Foreign-born persons are only slightly overrepresented in STEM jobs relative to all jobs in the economy; foreign-born workers represent 20 percent of the STEM workforce, as compared to 16 percent of the total U.S. workforce. However, they are greatly overrepresented in STEM jobs among persons with graduate degrees. This is most likely due to America's exemplary graduate schools that attract the best students from around the world, many of whom remain in the United States' workforce, enhancing America's competitiveness.

The aggregate statistics for these demographic groups mask some interesting variations seen within the groups. For example, Hispanics from different parts of North and South America exhibit quite different propensities to enter a STEM profession.

Echoing the findings of our earlier STEM reports, we find that STEM workers earn significantly more than comparable workers in non-STEM occupations, regardless of race, Hispanic origin, or nativity.

A common thread throughout this report and also touched upon in our earlier reports, is the importance of education as a gateway to high-quality STEM jobs. Enabling and encouraging equitable access to premium education is critical to ensuring that America maintains a wide and diverse source of STEM professionals that help to advance U.S. innovation and global competitiveness.

End Note

[1] Our earnings regression analyses control for many earnings-related characteristics, including age, gender, marital status, nativity and citizenship, educational attainment, region, major industry, and STEM occupation.

In: STEM Workforce Trends … ISBN: 978-1-62257-818-4
Editors: C. Martin Garcia and S. Cook

Chapter 5

TWO DECADES OF INCREASING DIVERSITY MORE THAN DOUBLED THE NUMBER OF MINORITY GRADUATE STUDENTS IN SCIENCE AND ENGINEERING*

Peter Einaudi[1]

From 1989 through 2009 the number of minority U.S. citizens and permanent residents enrolled in graduate science and engineering (S&E) programs more than doubled, growing from approximately 37,700 in 1989 to 92,700 in 2009. Increases in Hispanic, black, and Asian/Pacific Islander S&E graduate students were similar over this period (approximately 17,800, 18,200, and 17,200, respectively); however, these gains almost tripled the number of Hispanic graduate students (approximately 190% growth) and more than doubled the number of blacks (approximately 155% growth) and Asians/Pacific Islanders (approximately 110% growth). Enrollment among American Indians/Alaska Natives also nearly tripled, increasing from approximately 900 in 1989 to approximately 2,600 in 2009 (approximately 195% growth). Minority enrollment among U.S. citizens and permanent residents enrolled in graduate S&E programs grew from approximately 13% in 1989 to approximately 24% in 2009 (figure 1). Due to extra variability of the methodological changes in the 2007 Survey of Graduate Students and Postdoctorates in Science and Engineering (GSS), all growth rate calculations

* This is an edited, reformatted and augmented version of the National Center for Science and Engineering Statistics, NSF 11-319, dated July 2011.

comparing pre- and post-2007 counts are rounded to the nearest 5% and counts are rounded to the nearest 100; see "Data Limitations and Availability" for more information.

Despite these substantial gains, black and Hispanic U.S. citizens and permanent residents remain underrepresented within the S&E graduate student population when compared with the adult U.S. citizen population. In 2009, 7.8% of the U.S. citizens and permanent residents pursuing graduate S&E degrees were black and 7.1% were Hispanic. In 2009, 13.8% of U.S. citizens 21 to 45 years of age were black and 11.9% were Hispanic. In contrast, the percentage of American Indians/Alaska Natives in the 2009 S&E graduate student population was very similar to that of the adult U.S. citizen population (0.7% vs. 0.8%), and the percentage of Asians/Pacific Islanders pursuing S&E graduate degrees was more than twice that of the U.S. citizen population aged 21 to 45 years (8.6% vs. 3.8%).

These and other findings in this Info-Brief are from the fall 2009 GSS, cosponsored by the National Science Foundation (NSF) and the National Institutes of Health (NIH). The GSS is an annual survey of all academic institutions in the U.S. that grant research-based master's degrees or doctorates in science, engineering, or selected health (SEH) fields. The GSS collects data on the number and characteristics of graduate students, postdoctoral appointees (postdocs), and other doctorate-holding nonfaculty researchers in SEH fields.

Graduate Student Enrollment in S&E

In S&E fields, total graduate student enrollment (full-time and part-time) reached 545,685 in 2009, an increase of 3.1% from 2008 to 2009 (table 1). Graduate enrollment in engineering fields grew faster than in science fields for the fourth straight year (4.9% vs. 2.4% in 2009) and increased by approximately 40% over the past decade, as compared to approximately 30% for science fields. Although there was noted growth in S&E enrollment, assessing the overall trend in the selected health fields was not possible due to changes in the reporting of certain health fields. In particular, the 21.0% drop in the number of graduate students reported in "other health" fields in 2009 was primarily due to the exclusion of students in ineligible degree programs who had been included in prior years. Because these "other health" fields included ineligible students in prior years, the 2009 counts are not comparable to the 2008 and prior counts. Graduate enrollment within clinical health fields was fairly stable from 2008 to 2009 (0.8% growth) but had increased by approximately 45% over the prior 10 years.

The remainder of this InfoBrief focuses on the 2009 GSS data reported within S&E fields. Further analysis of GSS data on graduate enrollment in selected health fields can be obtained from NIH.[2]

Field of Study

From 2008 to 2009 the fastest growing graduate S&E fields were civil engineering, mechanical engineering, economics, biomedical engineering, political science, aerospace engineering, and agricultural sciences.[3] Of these fields, civil engineering experienced the fastest annual growth at 10.1%, but all grew by more than 7.0%. Over the 10-year span from 2000 to 2009, biomedical engineering easily outpaced all other S&E fields, increasing by approximately 145%.

Four fields declined from 2008 to 2009: psychology, sociology, history and philosophy of science, and sociology/anthropology; however, only for the combined field of sociology/anthropology did the data constitute a substantive downward trend. For sociology (as a single field) and history and philosophy of science, the declines were modest in number and were the first declines in several years; the 10-year trends for both fields showed increasing enrollment. As with the "other health" fields, much of the decline seen over the past 2 years in psychology was from the exclusion of graduate students in ineligible practitioner-oriented programs rather than declining enrollment within eligible research-oriented programs.

Enrollment Status

The S&E graduate students in 2009 were enrolled mostly as full-time students with only 27.0% enrolled part time (table 2).[4] Although there were more part-time S&E graduate students in 2009 than ever before, the proportion of part-time S&E graduate student enrollment has been in decline since 1985, and it dropped again in 2009 as growth in full-time enrollment outpaced that of part-time enrollment (3.9% to 1.0%). In contrast to part-time enrollment, first-time, full-time enrollment in S&E fields continued to rise in 2009, marking the fifth year in a row that the number of first-time, full-time graduate students hit a record high. In 2009 a total of 115,755 students started full-time graduate study in S&E programs, up 6.4% from 2008. First-time, full-time students accounted for 29.0% of all full-time graduate students in S&E and for 21.2% of all graduate students in S&E. Both are the highest proportions for first-time, full-time students since the mid-1980s.

Sex and Citizenship

Of the 115,755 first-time, full-time graduate students in S&E fields, 56.5% were men, contributing to a second straight year where men's overall enrollment in S&E fields grew faster than women's. In 2009 men's enrollment in S&E fields grew by 3.6% and women's enrollment grew by 2.5%. Given the relative stability from 2006 through 2009 of the proportion of S&E graduate students who were women, it seems that the long-standing trend toward greater gender parity among graduate students in S&E fields has been interrupted. From 1977 (the first year the GSS surveyed both master's- and doctorate-level academic institutions) through 2007, the proportion of S&E graduate students who were women increased every year and grew from 25.0% to 43.5%.

On the other hand, among the S&E graduate students with temporary visas, the gap between enrollment of women and men continued to narrow (women's enrollment increased by 3.4% vs. 1.9% for men from 2008 to 2009). However, temporary visa holders were much more likely than U.S. citizens and permanent residents to be men (65.7% vs. 52.5% in 2009). Temporary visa holders also remained much more likely than U.S. citizens or permanent residents to be enrolled full time (86.9% vs. 67.1% in 2009).

Despite a substantial change in the number of S&E graduate students holding temporary visas over the past 10 years, their proportion among all S&E graduate students has been relatively stable. From 2000 to 2009 approximately 3 out of every 10 graduate students in S&E were temporary visas holders. The percentage of S&E graduate students holding temporary visas was approximately 30% in 2000 and 2009, with a high of approximately 32% in 2002. The 2002 mark was the highest ever recorded in the GSS and reflected an increase of 8.7% over the 23.3% reported in 1995. Graduate enrollment of U.S. citizens and permanent residents grew at a slightly faster rate than that of foreign students on temporary visas from 2008 to 2009 (3.4% vs. 2.4%), marking the first time this has happened since foreign enrollment rebounded in 2005.

Table 1. Graduate enrollment in science, engineering, and health fields, by field: 2000–09

Field	2000	2001	2002	2003	2004	2005	2006	2007 old[a]	2007 new[a]	2008	2009	% change			
												2000–07	2007–09	2000–09	2008–09
All survey fields	493,311	509,607	540,404	567,121	574,463	582,226	597,643	607,823	619,499	631,489	631,645	23.2	2.0	30	0.0
Science and engineering	413,536	429,229	454,834	474,645	475,873	478,275	486,287	502,375	516,199	529,275	545,685	21.5	5.7	30	3.1
Science	309,424	319,736	335,166	347,268	352,307	357,710	363,246	372,120	384,523	391,419	401,008	20.3	4.3	30	2.4
Agricultural sciences	12,023	12,235	12,698	13,197	13,445	13,123	13,016	13,222	13,528	14,153	15,200	10.0	12.4	25	7.4
Biological sciences	56,282	57,639	61,088	64,701	66,565	68,479	69,941	71,663	71,932	72,666	73,304	27.3	1.9	30	0.9
Communication[a]	ne	ne	ne	ne	ne	ne	ne	ne	7,303	8,444	9,418	-	29.0	-	11.5
Computer sciences	47,350	52,196	55,269	53,696	50,016	47,978	47,653	48,959	48,246	49,553	51,161	3.4	6.0	10	3.2
Earth, atmospheric, and ocean sciences	13,941	13,841	14,240	14,620	15,131	14,836	14,920	14,675	14,100	14,389	14,839	5.3	5.2	5	3.1
Family and consumer sciences/human sciences[a]	ne	ne	ne	ne	ne	ne	ne	ne	2,780	3,549	3,794	-	36.5	-	6.9
Mathematical sciences	15,650	16,651	18,163	19,465	19,931	20,210	20,815	21,335	20,975	21,400	22,226	36.3	6.0	40	3.9
Multidisciplinary/inter disciplinary studies[a]	ne	ne	ne	ne	ne	ne	ne	ne	4,484	5,559	6,557	-	46.2	-	18.0
Multidisciplinary/inter disciplinary studies[a]	ne	ne	ne	ne	ne	ne	ne	ne	4,484	5,559	6,557	-	46.2	-	18.0
Neuroscience[a]	na	na	na	na	na	na	na	na	1,584	2,012	2,356	-	48.7	-	17.1

Table 1. (Continued)

Field	2000	2001	2002	2003	2004	2005	2006	2007 old[a]	2007 new[a]	2008	2009	% change			
												2000–07	2007–09	2000–09	2008–09
Multidisciplinary/interdisciplinary studies[a]	ne	ne	ne	ne	ne	ne	ne	ne	4,484	5,559	6,557	-	46.2	-	18.0
Neuroscience[a]	na	na	na	na	na	na	na	na	1,584	2,012	2,356	-	48.7	-	17.1
Physical sciences	30,385	31,038	32,341	34,298	35,761	36,375	36,901	37,111	36,824	37,319	38,149	22.1	3.6	25	2.2
Psychology	50,466	50,454	51,152	52,162	54,126	57,282	57,653	60,284	59,617	58,991	56,184	19.5	-5.8	10	-4.8
Social sciences	83,327	85,682	90,215	95,129	97,332	99,427	102,347	104,871	103,150	103,384	107,820	25.9	4.5	30	4.3
Agricultural economics	2,079	2,161	2,187	2,318	2,195	2,127	2,158	2,126	1,989	2,132	2,222	2.3	11.7	5	4.2
Anthropology	7,626	7,491	7,481	7,789	7,826	7,750	8,150	8,099	8,129	8,333	8,359	6.2	2.8	10	0.3
Economics	10,748	11,408	12,009	12,316	12,318	11,805	12,132	12,328	12,597	12,971	13,993	14.7	11.1	30	7.9
Geography	4,036	4,304	4,383	4,721	4,809	4,800	4,750	4,660	4,660	4,745	4,810	15.5	3.2	20	1.4
History and philosophy of science	532	571	663	737	994	965	968	1,119	1,054	1,177	1,006	110.3	-4.6	90	-14.5
Linguistics	2,674	2,744	2,875	3,028	2,941	3,187	3,074	3,076	2,879	3,095	3,170	15.0	10.1	20	2.4
Political science	31,131	31,805	34,934	36,880	39,023	40,780	41,784	41,854	41,349	40,871	43,919	34.4	6.2	40	7.5
Sociology	8,652	8,812	8,946	9,127	8,874	9,018	9,035	9,734	9,642	10,002	9,731	12.5	0.9	10	-2.7
Sociology/ anthropology	745	808	719	773	839	848	837	831	682	653	576	11.5	-15.5	-25	-11.8
Other social sciences	15,104	15,578	16,018	17,440	17,513	18,147	19,459	21,044	20,169	19,405	20,034	39.3	-0.7	35	3.2
Engineering	104,112	109,493	119,668	127,377	123,566	120,565	123,041	130,255	131,676	137,856	144,677	25.1	9.9	40	4.9
Aerospace engineering	3,407	3,451	3,685	4,048	4,089	4,170	4,482	4,616	4,616	4,902	5,266	35.5	14.1	55	7.4

Field	2000	2001	2002	2003	2004	2005	2006	2007 old[a]	2007 new[a]	2008	2009	% change			
												2000–07	2007–09	2000–09	2008–09
Engineering	104,112	109,493	119,668	127,377	123,566	120,565	123,041	130,255	131,676	137,856	144,677	25.1	9.9	40	4.9
Aerospace engineering	3,407	3,451	3,685	4,048	4,089	4,170	4,482	4,616	4,616	4,902	5,266	35.5	14.1	55	7.4
Architecture[a]	na	na	na	na	na	na	na	na	4,601	5,905	6,804	-	47.9	-	15.2
Biomedical engineering	3,197	3,599	4,338	5,301	5,807	6,067	6,482	6,881	6,904	7,339	7,904	115.2	14.5	145	7.7
Chemical engineering	7,056	6,913	7,414	7,516	7,452	7,173	7,261	7,383	7,584	7,892	8,188	4.6	8.0	15	3.8
Civil engineering[a]	16,451	16,665	17,713	18,890	18,561	18,114	17,802	19,867	16,071	16,931	18,638	20.8	16.0	15	10.1
Electrical engineering	33,611	36,100	39,948	41,763	38,995	37,450	38,265	40,207	40,588	41,164	41,218	19.6	1.6	25	0.1
Industrial engineering	12,119	12,940	14,033	14,313	13,852	13,650	13,829	14,290	14,474	15,692	15,825	17.9	9.3	30	0.8
Mechanical engineering	15,235	15,852	17,139	18,393	17,852	17,373	17,919	18,366	18,347	19,585	21,243	20.6	15.8	40	8.5
Metallurgical/materials engineering	4,377	4,721	4,992	5,131	5,059	5,160	5,268	5,365	5,314	5,539	5,863	22.6	10.3	35	5.8
Other engineering	8,659	9,252	10,406	12,022	11,899	11,408	11,733	13,280	13,177	12,907	13,728	53.4	4.2	60	6.4
Health[a]	79,775	80,378	85,570	92,476	98,590	103,951	111,356	105,448	103,300	102,214	85,960	32.2	-16.8	10	-15.9
Clinical medicine[c]	16,407	17,363	19,166	20,574	20,866	21,414	23,441	24,616	22,751	23,939	24,125	50.0	6.0	45	0.8

Table 1. (Continued)

Field	2000	2001	2002	2003	2004	2005	2006	2007 old[a]	2007 new[a]	2008	2009	% change			
												2000–07	2007–09	2000–09	2008–09
Other health[d]	63,368	63,015	66,404	71,902	77,724	82,537	87,915	80,832	80,549	78,275	61,835	27.6	-23.2	0	-21.0

na = not applicable; data were not collected at this level of detail. ne = not eligible; data were not collected for this field before 2007. - = not calculable.

[a] In 2007 the survey was redesigned and five fields were added or reclassified to improve reporting. "2007new" shows data as collected in 2007; "2007old" shows data as they would have been collected in prior years. See www.nsf.gov/statistics/nsf10307/ for more detail.

[b] "% change 2000–07" shows growth from 2000 to 2007old; "% change 2007–09" shows growth from 2007new to 2009; "% change 2000–09" shows growth from 2000 to 2009 and is rounded to nearest 5% to reflect the imprecision of this estimate due to inclusion of methodological changes.

[c] Clinical medicine includes research-oriented graduate students in these medical fields: anesthesiology, cardiology, endocrinology, gastroenterology, hematology, neurology, obstetrics/gynecology, oncology/cancer research, ophthalmology, otorhinolaryngology, pediatrics, preventive medicine/community health, psychiatry, pulmonary disease, radiology, surgery, and clinical medicine, not elsewhere classified.

[d] Counts within "other health" fields should not be used for trend analysis as counts prior to 2009 include significant numbers of ineligible students in practitioner-oriented degree programs. Starting in 2007, eligibility rules were clarified.

Source: National Science Foundation/National Center for Science and Engineering Statistics, NSF-NIH Survey of Graduate Students and Postdoctorates in Science and Engineering.

Table 2. Graduate enrollment in science, engineering, and health fields, by enrollment status, sex, citizenship, and race/ethnicity: 2000–09

Characteristic	2000	2001	2002	2003	2004	2005	2006	2007 old[a]	2007 new[a]	2008	2009	% change			
												2000 – 07	2007 – 09	2000 – 09	2008 – 09
All survey fields	493,311	509,607	540,404	567,121	574,463	582,226	597,643	607,823	619,499	631,489	631,645	23.2	2.0	30	0.0
Science and engineering	413,536	429,229	454,834	474,645	475,873	478,275	486,287	502,375	516,199	529,275	545,685	21.5	5.7	30	3.1
Full time	291,355	304,021	325,472	339,028	340,529	341,742	349,802	362,976	371,542	383,560	398,498	24.6	7.3	35	3.9
First time	78,332	82,411	86,827	89,331	86,565	89,038	94,413	98,205	100,990	108,819	115,755	25.4	14.6	50	6.4
Other	213,023	221,610	238,645	249,697	253,964	252,704	255,389	264,771	270,552	274,741	282,743	24.3	4.5	35	2.9
Part time	122,181	125,208	129,362	135,617	135,344	136,533	136,485	139,399	144,657	145,715	147,187	14.1	1.7	20	1.0
Male	243,057	251,810	266,217	276,248	274,008	271,967	275,181	284,080	288,926	297,278	307,935	16.9	6.6	25	3.6
Female	170,479	177,419	188,617	198,397	201,865	206,308	211,106	218,295	227,273	231,997	237,749	28.0	4.6	40	2.5
U.S. citizens and permanent residents															
S&E graduate enrollment	290,651	294,608	309,119	327,181	332,022	338,513	343,603	353,142	365,091	369,781	382,342	21.5	4.7	30	3.4
Full time	185,613	188,135	200,097	212,855	217,345	220,842	225,338	233,343	240,319	245,691	256,503	25.7	6.7	40	4.4
First time	46,301	48,207	54,625	59,649	58,853	60,157	60,978	62,009	64,284	68,093	75,321	33.9	17.2	65	10.6
Other	139,312	139,928	145,472	153,206	158,492	160,685	164,360	171,334	176,035	177,598	181,182	23.0	2.9	30	2.0
Part time	105,038	106,473	109,022	114,326	114,677	117,671	118,265	119,799	124,772	124,090	125,839	14.1	0.9	20	1.4
Male	156,975	157,945	164,891	174,818	176,297	177,900	179,783	184,498	188,642	191,989	200,641	17.5	6.4	30	4.5
Female	133,676	136,663	144,228	152,363	155,725	160,613	163,820	168,644	176,449	177,792	181,700	26.2	3.0	35	2.2
Hispanic	17,203	17,974	19,634	21,241	22,212	23,387	24,140	25,032	25,739	26,098	27,265	45.5	5.9	60	4.5
Non-Hispanic															
White	205,569	206,018	213,135	222,674	224,850	225,776	227,993	232,043	240,204	242,623	250,443	12.9	4.3	20	3.2
Asian/Pacific Islander	24,998	26,494	29,229	31,786	30,645	30,574	30,179	31,279	31,897	31,477	32,879	25.1	3.1	30	4.5

Table 2. (Continued)

Characteristic	2000	2001	2002	2003	2004	2005	2006	2007 old[a]	2007 new[a]	2008	2009	% change			
												2000–07	2007–09	2000–09	2008–09
Asian/Pacific Islander	24,998	26,494	29,229	31,786	30,645	30,574	30,179	31,279	31,897	31,477	32,879	25.1	3.1	30	4.5
Black	20,834	21,455	22,668	24,174	24,624	25,248	25,664	26,565	27,637	28,680	29,973	27.5	8.5	45	4.5
American Indian/															
Alaska Native	1,602	1,683	1,734	1,879	1,848	1,958	2,112	2,168	2,262	2,618	2,549	35.3	12.7	60	-2.6
Other or unknown race/ethnicity	20,445	20,984	22,719	25,427	27,843	31,570	33,515	36,055	37,352	38,285	39,233	76.4	5.0	90	2.5
Temporary visa holders															
S&E graduate enrollment	122,885	134,621	145,715	147,464	143,851	139,762	142,684	149,233	151,108	159,494	163,343	21.4	8.1	35	2.4
Full time	105,742	115,886	125,375	126,173	123,184	120,900	124,464	129,633	131,223	137,869	141,995	22.6	8.2	35	3.0
First time	32,031	34,204	32,202	29,682	27,712	28,881	33,435	36,196	36,706	40,726	40,434	13.0	10.2	25	-0.7
Other	73,711	81,682	93,173	96,491	95,472	92,019	91,029	93,437	94,517	97,143	101,561	26.8	7.5	40	4.5
Part time	17,143	18,735	20,340	21,291	20,667	18,862	18,220	19,600	19,885	21,625	21,348	14.3	7.4	25	-1.3
Male	86,082	93,865	101,326	101,430	97,711	94,067	95,398	99,582	100,284	105,289	107,294	15.7	7.0	25	1.9
Female	36,803	40,756	44,389	46,034	46,140	45,695	47,286	49,651	50,824	54,205	56,049	34.9	10.3	50	3.4
Health	79,775	80,378	85,570	92,476	98,590	103,951	111,356	105,448	103,300	102,214	85,960	32.2	-16.8	10	-15.9

S&E = science and engineering.

a In 2007 the survey was redesigned and five fields were added or reclassified to improve reporting. "2007new" shows data as collected in 2007; "2007old" shows data as they would have been collected in prior years. See appendix A in *Graduate Students and Postdoctorates in Science and Engineering: Fall 2007* (NSF 10-307 at www.nsf.gov/statistics/nsf10307/) for more detail. Caution should be used when calculating year-to-year growth and interpreting trends. b "% change 2000–07" shows growth from 2000 to 2007old; "% change 2007–09" shows growth from 2007new to 2009; "% change 2000–09" shows growth from 2000 to 2009 and is rounded to the nearest 5% to reflect the imprecision of this estimate due to the inclusion of methodological changes.

Note: The "Asian/Pacific Islander" category includes "Asian" and "Native Hawaiian/Other Pacific Islander"; the "Hispanic" category includes "Hispanic/Latino, one or more races"; the "Other or unknown" category includes "Non-Hispanic/Latino, more than one race" and "ethnicity/race unknown ornot stated."

Source: National Science Foundation/National Center for Science and Engineering Statistics, NSF-NIH Survey of Graduate Students and Postdoctorates in Science and Engineering.

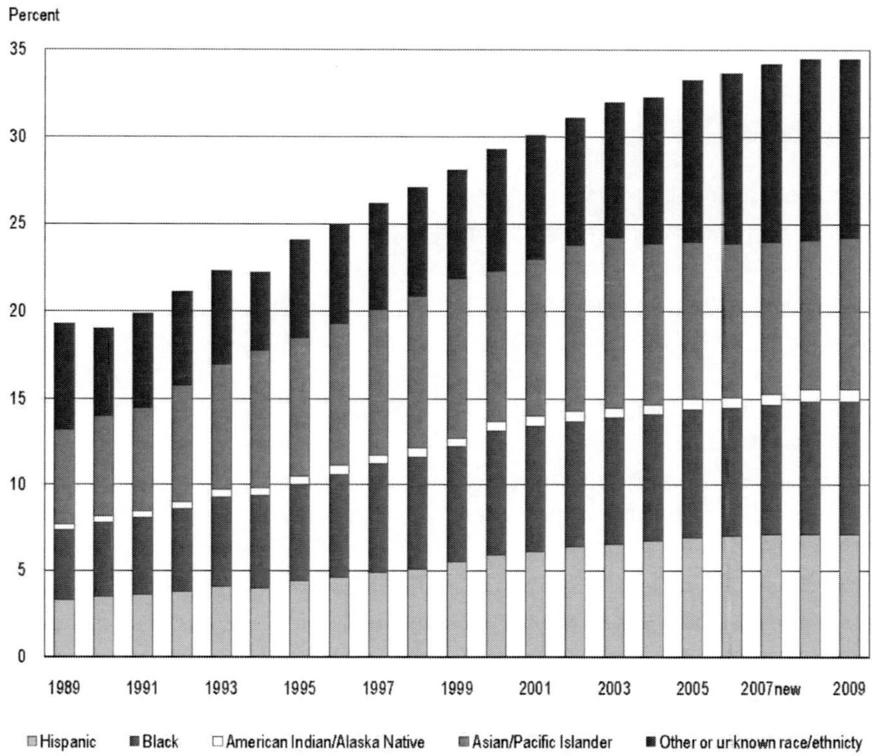

Notes: In 2007 the survey was redesigned and five fields were added or reclassified to improve reporting. "2007new" shows data as collected in 2007. See appendix A in *Graduate Students and Postdoctorates in Science and Engineering: Fall 2007* (NSF 10-307 at www.nsf.gov/statistics/nsf10307/) for more detail. Caution should be used when calculating year-to-year growth and interpreting trends. The "Asian/Pacific Islander" category includes "Asian" and "Native Hawaiian/Other Pacific Islander," the "Hispanic" category includes "Hispanic/Latino in one or more races," the "Other or unknown" category includes "Non-Hispanic/Latino in more than one race" and/or "ethnicity/race unknown or not stated."

Source: National Science Foundation/National Center for Science and Engineering Statistics, NSF-NIH Survey of Graduate Students and Postdoctorates in Science and Engineering.

Figure 1. Graduate enrollment among U.S. citizens and permanent residents in science and engineering fields, by minority and other/unknown race/ethnicity: 1989–2009.

Race and Ethnicity

From 1989 to 2009 the number of U.S. citizens and permanent residents from minority racial/ethnic groups enrolled in S&E graduate programs more than doubled, whereas the number of non-Hispanic white U.S. citizen or permanent resident graduate students increased by less than 10%.

As a result, the racial/ethnic makeup of U.S. citizens and permanent residents enrolled in S&E graduate programs (U.S. S&E graduate students) has become increasingly diverse (figure 1).

The growth in the number of Hispanic, black, and American Indian/Alaska Native S&E graduate students from 1989 to 2009 has been remarkably consistent, with increases every year for blacks, in 19 of 20 years for Hispanics, and in 17 of 20 years for American Indians/Alaskan Natives. As a result of two decades of growth, the number of graduate students who were Hispanic U.S. citizens or permanent residents nearly tripled, growing by approximately 190% (from approximately 9,400 in 1989 to approximately 27,300 in 2009).

S&E graduate enrollment among black U.S. citizens or permanent residents grew by approximately 155% (from approximately 11,800 to approximately 30,000) over this period, and American Indian/Alaska Native S&E graduate enrollment grew by approximately 195% (from approximately 900 to approximately 2,500).

S&E graduate enrollment among Asian/Pacific Islander U.S. citizens or permanent residents rose by approximately 110% over this period (from approximately 15,700 in 1989 to approximately 32,900 in 2009), but this growth was not as uniform as that among the other racial/ethnic minorities. In proportional terms, Hispanics increased from approximately 3% of all U.S. S&E graduate students in 1989 to approximately 7% in 2009, blacks increased from approximately 4% to approximately 8%, and American Indians/Alaskan Natives increased from less than 1% to approximately 1%.

In contrast, the proportion of Asians/Pacific Islanders varied over time: it grew from approximately 6% in 1989 to approximately 9% in 1999, dropped below this level briefly in 2000 and 2001, peaked at approximately 10% in 2003, and then declined to approximately 9% in 2008 and 2009.

U.S. S&E graduate students of other or unknown race/ethnicity also grew substantially from 1989 to 2009, with the majority of this growth occurring over the past decade. In 1989 there were approximately 17,200 graduate students reported as “Other or unknown” race, approximately 18,600 in 1999, and approximately 39,200 in 2009.

Since 2000 the "Other or unknown" race category has consisted of two collected categories: "More than one race (non-Hispanic/Latino)" and "Ethnicity/race unknown or not stated." In 2000 there were approximately 400 non-Hispanic graduate students with more than one race and approximately 20,000 graduate students with unknown or not stated race/ethnicity. In 2009 these figures were approximately 2,300 and 36,900. Since 2000 the "More than one race" category has grown much faster (approximately 425%) than the "Ethnicity/race unknown or not stated" category (approximately 85%); however, the vast majority of the S&E graduate students in the combined "Other or unknown" race category were those having unknown or not stated race/ethnicity.

First-Time, Full-Time Enrollment, by Citizenship and Field

In stark contrast to the substantial growth from 2008 to 2009 among U.S. citizens and permanent residents enrolling in full-time S&E programs for the first time (10.6%), the number of first-time, full-time graduate students with temporary visas declined (-0.7%) (table 3). This was the first decline in foreign first-time, full-time S&E enrollment since 2004. The year 2009 was the second year in a row that first-time, full-time enrollment in engineering grew faster among U.S. citizens and permanent residents (16.8%) than among foreign students with temporary visas (-1.9%).

Similar numbers of temporary visa holders enrolled in science fields for the first time in 2008 and 2009, but these students were distributed differently, with more entering the social and agricultural sciences in 2009 and fewer entering psychology and the computer, biological, and physical sciences.

POSTDOCTORAL APPOINTEES IN S&E

The GSS also collects information about postdocs who work at U.S. academic institutions (and their affiliates, such as research centers and hospitals) granting research-based master's or doctoral degrees in SEH fields. In total, 57,805 postdocs were reported in 2009, a 6.7% increase over 2008 (table 4). Over 70% of the postdocs were in S&E fields.

Table 3. First-time, full-time graduate enrollment in science, engineering, and health fields, by field and citizenship: 2005–09

Field	U.S. citizens and permanent residents							Temporary visa holders						
	2005	2006	2007 old[a]	2007 new[a]	2008	2009	% change 2008–09	2005	2006	2007 old[a]	2007 new[a]	2008	2009	% change 2008–09
All survey fields	79,531	81,142	82,127	83,932	87,812	92,275	5.1	30,688	35,340	38,109	38,517	42,823	42,481	-0.8
Science and engineering	60,157	60,978	62,009	64,284	68,093	75,321	10.6	28,881	33,435	36,196	36,706	40,726	40,434	-0.7
Science	49,252	49,231	50,005	52,017	54,748	59,733	9.1	17,413	19,306	20,382	20,708	23,479	23,519	0.2
Agricultural sciences	1,828	1,764	1,837	1,848	1,996	2,359	18.2	432	442	476	508	601	635	5.7
Biological sciences	9,925	9,946	10,230	10,507	10,870	11,825	8.8	2,993	3,109	3,255	3,242	3,658	3,618	-1.1
Communication[a]	ne	ne	ne	1,229	1,489	1,677	12.6	ne	ne	ne	302	423	442	4.5
Computer sciences	3,632	3,382	3,077	3,034	3,324	3,659	10.1	4,626	5,601	6,275	6,222	7,110	6,898	-3.0
Earth, atmospheric, and ocean sciences	2,270	2,289	2,364	2,271	2,306	2,533	9.8	488	506	497	478	598	620	3.7
Family and consumer sciences/human sciences[a]	ne	ne	ne	456	515	576	11.8	ne	ne	ne	36	64	57	-10.9
Mathematical sciences	2,561	2,522	2,629	2,561	2,548	2,808	10.2	1,635	1,820	1,966	1,871	2,043	2,080	1.8
Multidisciplinary/ interdisciplinary studies[a]	ne	ne	ne	584	859	1,131	31.7	ne	ne	ne	187	281	281	0.0
Neuroscience[a]	na	na	na	180	259	334	29.0	na	na	na	93	45	61	35.6
Physical sciences	4,138	4,045	4,146	4,089	3,988	4,424	10.9	2,519	2,581	2,641	2,622	2,904	2,852	-1.8
Psychology	9,800	9,645	9,985	9,861	10,407	10,347	-0.6	558	579	639	615	675	633	-6.2

Characteristic	2000	2001	2002	2003	2004	2005	2006	2007 old[a]	2007 new[a]	2008	2009	% change[b]			
												2000 – 07	2007 – 09	2000 – 09	2008 – 09
Industrial engineering	48	21	43	45	50	51	51	73	71	115	109	52.1	53.5	125	-5.2
Mechanical engineering	480	501	441	543	514	562	644	725	722	784	948	51.0	31.3	100	20.9
Metallurgical/materials															
engineering	507	479	507	539	567	578	571	555	564	605	758	9.5	34.4	50	25.3
Other engineering	424	483	438	522	596	570	706	675	671	751	807	59.2	20.3	90	7.5
Health[a]	12,891	13,115	13,097	13,062	13,175	14,099	14,456	14,818	14,617	15,961	17,001	14.9	16.3	30	6.5

na = not applicable; data were not collected at this level of detail. ne = not eligible; data were not collected for this field before 2007. - = not calculable.

a In 2007 the survey was redesigned and five fields were added or reclassified to improve reporting. “2007new” shows data as collected in 2007; “2007old” shows data as they would have been collected in prior years. See appendix A in *Graduate Students and Postdoctorates in Science and Engineering: Fall 2007* (NSF 10-307 at www.nsf.gov/statistics/nsf10307/) for more detail. Caution should be used when calculating year-to-year growth and interpreting trends.

b “% change 2000–07” shows growth from 2000 to 2007old; “% change 2007–09” shows growth from 2007new to 2009; “% change 2000–09” shows growth from 2000 to 2009 and is rounded to the nearest 5% to reflect the imprecision of this estimate due to the inclusion of methodological changes.

Source: National Science Foundation/National Center for Science and Engineering Statistics, NSF-NIH Survey of Graduate Students and Postdoctorates in Science and Engineering.

Table 4. (Continued)

Characteristic	2000	2001	2002	2003	2004	2005	2006	2007 old[a]	2007 new[a]	2008	2009	% change[b]			
												2000 – 07	2007 – 09	2000 – 09	2008 – 09
Agricultural sciences	822	833	963	1,054	959	1,007	927	948	985	1,147	1,083	15.3	9.9	30	-5.6
Biological sciences	16,734	17,032	17,640	18,625	18,716	18,747	18,807	19,218	19,109	19,827	20,159	14.8	5.5	20	1.7
Communication[a]	ne	ne	ne	ne	ne	ne	ne	ne	30	32	38	-	26.7	-	18.8
Computer sciences	344	336	356	355	384	406	467	516	456	493	594	50.0	30.3	75	20.5
Earth, atmospheric, and ocean sciences	1,155	1,049	1,129	1,182	1,263	1,364	1,495	1,322	1,250	1,339	1,424	14.5	13.9	25	6.3
Family and consumer sciences/human sciences[a]	ne	ne	ne	ne	ne	ne	ne	ne	8	19	22	-	175.0	-	15.8
Mathematical sciences	385	353	395	449	468	500	579	621	624	723	737	61.3	18.1	90	1.9
Multidisciplinary/ interdisciplinary studies[a]	ne	ne	ne	ne	ne	ne	ne	ne	244	348	459	-	88.1	-	31.9
Neuroscience[a]	na	na	na	na	na	na	na	na	285	343	645	-	126.3	-	88.0
Physical sciences	6,270	6,223	6,619	6,829	7,059	7,011	6,703	6,760	6,719	6,885	7,447	7.8	10.8	20	8.2
Psychology	730	809	815	960	902	884	873	1,106	1,088	1,077	1,219	51.5	12.0	65	13.2
Social sciences	471	409	454	402	365	371	394	495	483	508	561	5.1	16.1	20	10.4
Engineering	3,313	3,152	3,566	3,810	3,949	4,166	4,642	4,908	4,942	5,462	6,416	48.1	29.8	95	17.5
Aerospace engineering	111	128	140	141	141	153	165	178	178	154	168	60.4	-5.6	50	9.1
Architecture[a]	na	na	na	na	na	na	na	na	5	11	22	-	340.0	-	100.0
Biomedical engineering	220	262	284	388	425	477	591	640	640	710	960	190.9	50.0	335	35.2
Chemical engineering	703	574	758	686	689	702	735	758	790	880	1,084	7.8	37.2	55	23.2
Civil engineering[a]	295	268	342	300	313	384	458	419	417	465	535	42.0	28.3	80	15.1
Electrical engineering	525	436	613	646	654	689	721	885	884	987	1,025	68.6	16.0	95	3.9

Field	U.S. citizens and permanent residents							Temporary visa holders						
	2005	2006	2007 old[a]	2007 new[a]	2008	2009	% change 2008–09	2005	2006	2007 old[a]	2007 new[a]	2008	2009	% change 2008–09
Social sciences	15,098	15,638	15,737	15,397	16,187	18,060	11.6	4,162	4,668	4,633	4,532	5,077	5,342	5.2
Engineering	10,905	11,747	12,004	12,267	13,345	15,588	16.8	11,468	14,129	15,814	15,998	17,247	16,915	-1.9
Health[a]	19,374	20,164	20,118	19,648	19,719	16,954	-14.0	1,807	1,905	1,913	1,811	2,097	2,047	-2.4

na = not applicable; data were not collected at this level of detail. ne = not eligible; data were not collected for this field before 2007.

a In 2007 the survey was redesigned and five fields were added or reclassified to improve reporting. “2007new” shows data as collected in 2007; “2007old” shows data as they would have been collected in prior years. See appendix A in *Graduate Students and Postdoctorates in Science and Engineering: Fall 2007* (NSF 10-307 at www.nsf.gov/statistics/nsf10307/) for more detail. Caution should be used when calculating year-to-year growth and interpreting trends.

Source: National Science Foundation/National Center for Science and Engineering Statistics, NSF-NIH Survey of Graduate Students and Postdoctorates in Science and Engineering.

Table 4. Postdoctoral appointees in science, engineering, and health fields, by sex, citizenship, and field: 2000–09

Characteristic	2000	2001	2002	2003	2004	2005	2006	2007 old[a]	2007 new[a]	2008	2009	% change[b]			
												2000 – 07	2007 – 09	2000 – 09	2008 – 09
All survey fields	43,115	43,311	45,034	46,728	47,240	48,555	49,343	50,712	50,840	54,164	57,805	17.6	13.7	35	6.7
Science and engineering	30,224	30,196	31,937	33,666	34,065	34,456	34,887	35,894	36,223	38,203	40,804	18.8	12.6	35	6.8
Male	21,296	20,941	21,807	22,882	23,080	23,227	23,361	24,412	24,631	25,119	26,647	14.6	8.2	25	6.1
Female	8,928	9,255	10,130	10,784	10,985	11,229	11,526	11,482	11,592	13,084	14,157	28.6	22.1	60	8.2
U.S. citizens and permanent residents	12,627	12,073	13,524	13,542	13,969	14,078	14,111	14,903	15,107	16,274	18,175	18.0	20.3	45	11.7
Temporary visa holders	17,597	18,123	18,413	20,124	20,096	20,378	20,776	20,991	21,116	21,929	22,629	19.3	7.2	30	3.2
Science	26,911	27,044	28,371	29,856	30,116	30,290	30,245	30,986	31,281	32,741	34,388	15.1	9.9	30	5.0

Field of Appointment

Although most S&E postdocs work in science fields (84.3% in 2009), the proportion has declined each year since 2001 (when it was 89.6%), reflecting increasing numbers of postdocs working in engineering fields. From 2000 to 2009 the growth of engineering postdocs (approximately 95%) substantially outpaced that of science postdocs (approximately 30%). Single-year growth rates for engineering postdocs were 10.5% in 2008 and 17.5% in 2009.

Within each engineering field, postdoc employment grew substantially from 2000 to 2009. Mimicking the trend seen among graduate students, biomedical engineering was by far the fastest growing postdoc field over the past decade, increasing by approximately 335% from 2000 to 2009. The fastest growing science fields for postdocs from 2000 to 2009 were mathematical sciences (approximately 90% growth), computer sciences (approximately 75% growth), and psychology (approximately 65% growth).

Demographic Characteristics

In 2009, 65.3% of postdocs in S&E were men and 55.5% were temporary U.S. visa holders. The gap in the gender composition of S&E postdocs narrowed slightly from 2008 to 2009 as women's appointments grew at a slightly faster rate than men's (8.2% vs. 6.1%). The percentage of S&E postdocs who were women increased from 34.2% in 2008 to 34.7% in 2009. Over the past decade, the number of female postdocs increased by approximately 60%.

For the third straight year, growth in appointments of U.S. citizens and permanent residents outpaced that for temporary visa holders, increasing by 11.7% from 2008 to 2009 as compared with 3.2% for temporary visa holders. As a result, the percentage of postdocs who were U.S. citizens or permanent residents (44.5%) rose to the highest level since 1998, when it was 46.5%.

As seen in figure 2, the proportion of postdocs with temporary visas is correlated to the proportion of graduate students with temporary visas, although a higher percentage of postdocs than graduate students hold temporary visas. The percentage of foreign graduate students declined each year from 2002 to 2005, but an increasing enrollment of foreign graduate students since 2005 may suggest that the decreases seen in the percentage of foreign postdocs could level off and begin to rise over the next few years.

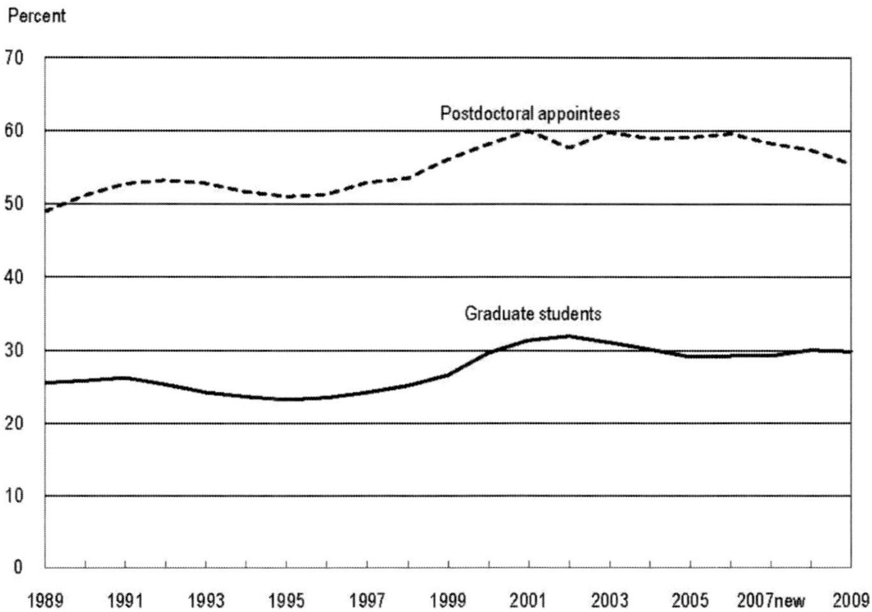

Notes: In 2007 the survey was redesigned and five fields were added or reclassified to improve reporting. "2007new" shows data as collected in 2007. See appendix A in *Graduate Students and Postdoctorates in Science and Engineering: Fall 2007* (NSF 10-307 at www.nsf.gov/statistics/nsf10307/) for more detail. Caution should be used when calculating year-to-year growth and interpreting trends.

Source: National Science Foundation/National Center for Science and Engineering Statistics, NSF-NIH Survey of Graduate Students and Postdoctorates in Science and Engineering.

Figure 2. Graduate students and postdoctoral appointees in science and engineering fields with temporary visas: 1989–2009.

SURVEY INFORMATION

During the production of this InfoBrief the America COMPETES Reauthorization Act of 2010 was signed into law. Section 505 of the bill renames the Division of Science Resources Statistics as the National Center for Science and Engineering Statistics (NCSES). The Center retains its reporting line to the Directorate for Social, Behavioral and Economic Sciences within the National Science Foundation. The new name signals the central role of NCSES in the collection, interpretation, analysis, and dissemination of objective data on the science and engineering enterprise.

Conducted since 1966, the GSS is an annual survey of all academic institutions in the U.S. granting research-based master's or doctoral degrees in SEH fields. The 2009 GSS collected data from 13,285 organizational units (departments, programs, affiliated research centers, and health care facilities) at 575 institutions of higher education and their affiliates in the United States, Puerto Rico, and Guam. The institutional response rate was 99.3%. An overview of the survey objectives and design can be found at http://www.nsf.gov/statistics/srvygradpostdoc/.

DATA LIMITATIONS AND AVAILABILITY

This publication provides the first release of data from the fall 2009 cycle of the GSS. The full set of detailed statistical tables from this survey will be available in the forthcoming report *Graduate Students and Postdoctorates in Science and Engineering: Fall 2009* at http://www.nsf.gov/statistics/grad-postdoc/.[5] Individual detailed tables may be available upon request in advance of the full report by contacting Kelly Kang (kkang@nsf.gov, 703-292-7796).

The GSS data from 1972 through 2009 are available in public-use format on the NSF website (http://www.nsf.gov/statistics/srvygradpostdoc/pub_data.cfm) and from the WebCASPAR data system (http://webcaspar.nsf.gov/).

Due to methodological changes in 2007, the data collected from 2007 through 2009 are not strictly comparable to those collected prior to 2007. As a result, care should be used when assessing trends within the GSS data. In this InfoBrief, the "2007new" column reports the data as collected in 2007 and the "2007old" column provides data as they would have been collected in 2006. Ten-year trends reported in the tables are labeled "% change 2000–09." Note that these percentages are rounded to the nearest 5% to reflect the extra variability in the estimate because of the methodological change that occurred in 2007. To help readers understand trends in light of the change in 2007, two additional columns are included in the tables. The "percentage change 2000–07" reflects the growth from 2000 to 2007old and the "percentage change 2007–09" reflects the growth seen from 2007new to 2009. Please see appendix A, "Technical Notes," in *Graduate Students and Postdoctorates in Science and Engineering: Fall 2007* (NSF 10-307) for a more detailed discussion of these changes.

Estimates for the adult U.S. citizen and permanent resident population are based on the 2009 American Community Survey (ACS) conducted by the U.S. Census Bureau (http://usa.ipums.org/usa/sda/).

Age was limited to 21–45 in order to better estimate the graduate student population.

The GSS collects data on graduate students, postdocs, and other doctorate-holding nonfaculty researchers in research-oriented science, engineering, and health fields. Practitioner-oriented degrees within these fields (e.g., master's degrees in nursing and physical therapy) are not eligible for the GSS. Although these degrees have never been eligible in GSS, many respondents were not aware of this. After these exclusions were made more explicit in 2007, fewer graduate students in practitioner-oriented degree programs have been reported in the past three survey cycles. The large drop from 2008 to 2009 within the "other health" fields resulted from an increased effort to eliminate ineligible students from the GSS. The majority of excluded degrees are located within these "other health" fields. The GSS Field Code List provides a complete list of eligible and ineligible fields (http://www.nsf.gov/statistics/nsf10307/pdf/Complete_GSS_Code_List.pdf).

End Notes

[1] Peter Einaudi, research analyst, RTI International, 3040 Cornwallis Road, P.O. Box 12194, Research Triangle Park, NC 27709-2194.

[2] The GSS collects data on health fields selected by NIH. These fields make up about one-third of all health fields in the U.S. Department of Education Classification of Instructional Programs (CIP) taxonomy. The majority of excluded degrees are in practitioner-oriented fields that do not meet the research-based criteria for GSS eligibility. NIH information on trends seen within selected health fields can be found at http://www.report.nih.gov/nihdatabook/Default.aspx?catid=19.

[3] The five fields (communication, family and consumer sciences/human sciences, multidisciplinary/interdisciplinary, architecture, neuroscience) first reported in 2007 also grew quickly; however, these trends should be treated with caution given the lack of comparable data. High growth rates may also be due to increasing awareness of these fields' eligibility and therefore increased reporting of these fields, rather than growth within the field.

[4] Full-time enrollment is defined according to the institution's policies and definition. First-time graduate students are those enrolled for graduate credit for the first time as of fall 2009 at the institution at which they are pursuing a degree.

[5] For further information on the survey, please contact Kelly H. Kang, Human Resources Statistics Program, National Center for Science and Engineering Statistics, National Science Foundation, 4201 Wilson Boulevard, Suite 965, Arlington, VA 22230 (kkang@nsf.gov, 703-292-7796).

In: STEM Workforce Trends … ISBN: 978-1-62257-818-4
Editors: C. Martin Garcia and S. Cook

Chapter 6

TESTIMONY OF DARLA WHITAKER, SENIOR VICE PRESIDENT, WORLDWIDE HUMAN RESOURCES, TEXAS INSTRUMENTS. HEARING ON "STEM THE TIDE: SHOULD AMERICA TRY TO PREVENT AN EXODUS OF FOREIGN GRADUATES OF U.S. UNIVERSITIES WITH ADVANCED SCIENCE DEGREES?"*

Chairman Gallegly, Ranking Member Lofgren, Chairman Smith, Ranking Member Conyers, Members of the subcommittee, I am Darla Whitaker, Senior Vice President for Worldwide Human Resources at Texas Instruments. I am also representing the views of the Semiconductor Industry Association which is a member of Compete America. Thank you for inviting me to speak to you today about how best to retain the talent of U.S. university graduates holding advanced degrees in Science, Technology, Engineering and Math (STEM) fields.

Over the course of my testimony I plan to tell you a bit about TI, our hiring practices and needs and suggest some ways in which Congress can help ensure that companies like TI can have access to the best minds emerging from US universities.

---

* This is an edited, reformatted and augmented version of the testimony given on October 5, 2011 before the House Committee on the Judiciary, Subcommittee on Immigration.

About Texas Instruments

While many people think of calculators when they think of TI, our primary business is to design and manufacture semiconductors –or chips. We are, in fact, the world's 3rd largest semiconductor company. Chips are the engines, "the building blocks" if you will, of all electronic devices and machines. Semiconductors have fundamentally changed the way people around the globe work, learn and play. They have driven unprecedented productivity across every sector of the economy and have enabled the development of numerous new industries over the last 50 years.

Chips are in every electronic product, including computers, appliances and automobiles, agricultural machinery and aircraft. They are the "smart" in smart grids, appliances, and smart phones. We are driving solutions to some of the world's most pressing challenges in energy efficiency, heath care and national and homeland security. Texas Instruments is a global company, with manufacturing, design and sales operations in more than 30 countries and approximately 34,500 employees worldwide. In Texas, we have 10,400 employees. And with our recent acquisition of National Semiconductor, we now have a bigger footprint in Silicon Valley, as well.

Innovation is the cornerstone of our company. Founded in the 1930s as a geophysical exploration company that took used seismic signal processing technology to detect oil deposits, we have reinvented ourselves numerous times over the course of our history.

Over the last three years alone, we've invested $5 billion in research and development. TI's comprehensive innovation strategy includes funding and collaborating with universities and industry consortia, incubating breakthrough ideas in our own Kilby Labs, executing competitive roadmaps within TI's business units and specialized labs, and developing world-class manufacturing technologies. Texas Instruments has over 60,000 products and releases about 900 new products per year. Our engineers have developed more than 38,000 patents issued worldwide. More than 1,200 of those patents were issued in 2010.

Hiring at Texas Instruments

Innovation requires innovators. TI is fundamentally a company of engineers and scientists. Electrical engineers, in particular, are the life blood of our industry whether they are designing, manufacturing or selling our products.

To find those innovators, TI recruits heavily at top US engineering universities. We recruit the best engineering students, looking not only at their grades, but also to their creativity, community involvement and leadership skills.

The competition for these graduates is stiff. Most of the students to whom we make offers have other job offers, often from our competitors.

To find the best students, we work hard to develop relationships with top engineering universities. We build ties with student organizations, attend campus career fairs, and offer highly competitive internships, and support university research in partnership with Federal and State funding. We have an active co-op program that provides summer jobs to engineering students to help us identify top talent early.

TI has recently increased the number of engineering interns by nearly 60 percent. These candidates get to stand side-by-side with TI's best engineers and innovators and do hands-on, real world engineering. Our internship program also gives TI the chance to evaluate these rising talents and see if they have the creativity to be TI innovators. Currently about 40 percent of our new engineering hires now come from our internship program and we hope to move that up to 60-75 percent.

Texas Instruments' goal is to hire the best engineers and innovators from U.S. universities and to retain them. We do not choose where those engineers were born or what their citizenship is. We choose the best, the brightest and the most creative engineering graduates.

But as this Committee knows, in the past two decades we have seen some alarming trends. While the vast majority of BSEEs graduating from U.S. universities are American citizens, the numbers are significantly different at the graduate level.

The majority of those graduating from U.S. universities with advanced degrees in electrical engineering (EE) are foreign nationals. Of EEs graduating from U.S. universities with Master's degrees, 55% are foreign nationals. Of PhDs, 63% are.

TI doesn't choose the pool of graduates, we recruit from it.

Perhaps most significantly, while unemployment is high generally in the U.S., in engineering, it is not. Recent Labor Department statistics place the unemployment rate for electrical and electronics engineers at 3.7%. The competition for STEM talent is tight.

WHAT CONGRESS CAN DO

TI hires at all spectrums of the university engineering education continuum – from associates at community colleges to PhDs. It is for these upper level degrees that we have regularly found ourselves coming to Congress to seek help in enabling us to hire this highly educated talent.

We want these individuals to join our company, not on a temporary basis, but as permanent employees to provide long term value to our shareholders, customers and community. The current system in which they must first obtain an H-1B visa and then wait for years – sometimes up to a decade for a green card is frustrating for them, limits employer flexibility, and diminishes productivity. While our employees wait – and wait – for their green cards, their ability to be promoted or change jobs is limited.

The current immigration system allocates insufficient numbers to allow engineers and innovators to get green cards in a reasonable amount of time.

The current system also places an arbitrary cap that limits the number of immigrants from any one country to seven percent of the total number of immigrant visas issued each year. This "per country limit" has led to long wait times for our employees – reaching to more than nine years for Indian nationals in the third employment preference.

This is not sustainable. It hurts our company and our industry. It places unnecessary burdens and stresses on our employees. It harms American competiveness, even as other countries move to provide easier paths to permanent residence for STEM graduates.

It is also easily fixable.

We believe that a more direct path to obtaining a green card is optimal for these advanced degree innovators. A bill to increase the number of green cards available to graduates of U.S. universities holding advanced degrees in STEM would alleviate the long backlogs our innovators are stuck in, and would allow TI and the semiconductor industry to recruit and retain top talent more competitively.

It is imperative that the increase include not only PhD graduates, but Master's STEM graduates, as well. MSEE's are critical to innovation. While more PhDs may have their names on our 38,000 patents, much of the work is done in collaboration with our engineers with Master's degrees. And others are producing patents on their own.

For example, Sameer Pendharkar, a TI Fellow in our Analog Technology Development group has a Masters in electrical engineering and has produced over 50 patents for TI. Two years ago, he was recognized by The Academy of

Medicine, Engineering, and Science of Texas (TAMEST) as the recipient of their prestigious Edith and Peter O'Donnell Award, established to acknowledge achievements by young researchers in these disciplines. We have many other examples.

Among our foreign nationals who would qualify under a STEM bill for employees who graduate from U.S. universities with advanced degrees, the ratio of Master's level engineers to PhD engineers is at least four to one.

By modestly increasing green card numbers to allow employers to sponsor graduates of U.S. universities holding advanced degrees in STEM fields, Congress would vastly improve American competitiveness and secure our place as the world's innovation leader. This narrowly drawn, targeted approach will give employers like TI a more effective, timely way to sponsor advanced STEM graduates so they can apply their skills directly, here in the US, to innovating and creating a better future.

We also support H.R. 3012, the Fairness for High-Skilled Immigrants Act, which was recently introduced by Representative Chaffetz. That bill would eliminate the per country limit over a three year period and help rationalize the green card system.

We strongly encourage the Congress to urgently take up and pass legislation that would allow employers like Texas Instruments to more competitively recruit and retain graduates of U.S. universities holding advanced STEM degrees.

While we seek these changes, TI is very focused on increasing the pipeline of American born students receiving engineering degrees. That is why we are so passionate about university research funding and STEM education.

STEM education is our top philanthropic priority. In the past five years we've invested more than $150 million through TI and the TI foundation to support education at the K-12 and university levels. In addition, TI's employees invested countless volunteer hours in advancing STEM education in local schools and civic organizations. Our aim is to build an ecosystem of innovation by supporting student achievement, teacher effectiveness, and diversity, in STEM.

Among the initiatives we've undertaken is "Visioneering," a unique, annual event held at Southern Methodist University (SMU) during National Engineers Week. Since 2001, Visioneering has brought together approximately 9,000 middle school students and teachers, working engineers and innovators to explore how and why engineering makes a difference in the world around us.

We've also supported The Advanced Placement Incentive program that has had extraordinary results in increasing the number of underrepresented students taking and passing AP math and science tests, the UTEACH program that trains effective math and science teachers, the National Society of Black Engineers (NSBE), the Urban League, and, through the Women of TI Fund, High-Tech High Heels, to support and increase gender equity and diversity, along with innovation in STEM fields.

The point is this, engineering, math and science skills are critical to innovating and competing globally, and we need a well-educated technical workforce. I have provided a more complete listing of our activities in this area for the record.

At TI, we know that, to be successful and compete in our global market, we need the best, brightest and most innovative engineers. It's the foundation of our success and the success of the entire semiconductor industry.

Thank you for your time and attention, and for the opportunity to speak today on this important subject. I look forward to answering any questions you may have.

In: STEM Workforce Trends … ISBN: 978-1-62257-818-4
Editors: C. Martin Garcia and S. Cook

Chapter 7

STATEMENT OF VIVEK WADHWA, DIRECTOR OF RESEARCH, CENTER FOR ENTREPRENEURSHIP AND RESEARCH COMMERCIALIZATION, DUKE UNIVERSITY. HEARING ON "STEM THE TIDE: SHOULD AMERICA TRY TO PREVENT AN EXODUS OF FOREIGN GRADUATES OF U.S. UNIVERSITIES WITH ADVANCED SCIENCE DEGREES?"*

Subcommittee Chairman Elton Gallegly and ranking member Zoe Lofgren, Committee Chairman Lamar Smith and ranking member John Conyers, and distinguished members of the subcommittee, thank you for the opportunity to testify today.

I am an immigrant who came to the U.S. in 1980, to study. I ended up founding two software companies that created jobs for hundreds of American workers and that helped improve the productivity of many American businesses. Then, as a way of giving back to this great country for the opportunities it has given me, I became an academic. I now teach at several universities, conduct academic research on U.S. competitiveness, and share

* This is an edited, reformatted and augmented version of the testimony given on October 5, 2011 before the House Committee on the Judiciary, Subcommittee on Immigration.

my ideas through my *Washington Post* and *Bloomberg BusinessWeek* columns.

What I learned since becoming an academic is that the world has changed much faster than academics and policy makers understand.

FOREIGN STUDENTS' BELIEFS AND INTENTIONS

The week of September 19, I taught classes at Tsinghua University, in Beijing, China, for an entrepreneurship program run by UC-Berkeley's Center for Entrepreneurship, and I met local entrepreneurs at a local technology incubator.

The students there were very much like my students here—smart, ambitious, and open minded—but even more hungry for knowledge, more passionate about completing advanced degrees, and more motivated to become entrepreneurs. They were very eager to come to the United States to study. They saw education as the best path from poverty to prosperity. Entrepreneurship, for these students, was a way to rise above "the system" and be their own bosses. It is an opportunity to customize a path to success. The reason Tsinghua University, which is considered to be China's Harvard, spends hundreds of thousands of dollars to bring in lecturers from UC-Berkeley is that they know that American education is the best in the world—that it is what gives America its superior advantage in innovation and competitiveness.

But, unlike previous generations of Chinese students, the Tsinghua students didn't plan to come and stay in the U.S. They planned to take their knowledge back to China—where they are wanted. Most would readily start their companies in Silicon Valley or work in the U.S. after they graduate. But all have heard horror stories from their friends about the challenges that foreign students in the U.S. face in getting visas and jobs, so they won't even try. They know that many of America's leading companies have stopped interviewing foreign students because it's hard to obtain visas and because they may face a backlash for hiring foreigners. Given this, they see better opportunities in China and have no reason to consider staying in America.

This is consistent with the trend in the U.S.

I joined the Masters of Engineering Management program in the Pratt School of Engineering at Duke University in 2005. When I asked foreign students in the graduating class whether they planned to stay permanently in the U.S., the vast majority said they did. A few said they wanted to work in the

U.S. for a few years before deciding whether to make America their home. I have been asking the same question of my students every year since then. Now students ask what I mean by "permanently," or they ask why. It is now customary for students to seek a one- or two-year internship to gain U.S. work experience before heading home. Students here have heard horror stories from their predecessors similar to the ones told to the students I taught in China. They start looking for opportunities in their home countries well before they graduate.

To validate the anecdotal data we had gathered, my research team at Duke, Harvard, and UC-Berkeley surveyed 1,224 foreign nationals who were studying in U.S. institutions of higher learning or who had graduated by the end of the 2008 academic school year. We published our findings through the Kauffman Foundation, in a report titled *Losing the World's Best and Brightest: America's New Immigrant Entrepreneurs, Part V*. We confirmed that very few foreign students now plan to stay in the United States permanently—only six percent of Indian, 10 percent of Chinese, and 15 percent of European students. Here are the most important findings:

- A leading reason for students' intentions to depart is the fear that they will not be able to find a job in the United States upon graduation —a fear fuelled by their growing belief that the U.S. economy will lag behind average global growth rates.
- A significant majority of foreign students—85 percent of Indian and Chinese and 72 percent of European—are concerned about obtaining work visas. And 74 percent of Indian, 76 percent of Chinese, and 58 percent of European students are worried about obtaining jobs in their fields.
- Chinese students, in particular, strongly feel that the best employment prospects lie in their home country. Fifty-two percent (in comparison with 32 percent of Indian and 26 percent of Europeans) said that their home country offered the best job opportunities. This contrasts starkly with the belief held by a majority of skilled immigrants in the '80s and '90s that the best opportunities were in the U.S.
- Most foreign students are more optimistic about their home countries' economic future than the United States'. Whereas 7 percent of Chinese students, nine percent of European students, and 25 percent of Indian students stated that they believe the best days of the U.S. economy lie ahead, 74 percent of Chinese students and 86 percent of

Indian students felt that the best days for their home countries' economy lie ahead.

- Most have entrepreneurial hopes: 64 percent of Indian, 66 percent of European, and 68 percent of Chinese students indicated that they want to start a business within the next decade. For Indian and Chinese students, the majority (53 percent and 55 percent respectively) hope to start businesses in their home countries. Only 35 percent of European students wish to open a business in their home country.

What does this mean? It means the world's best and brightest aren't begging to be let into the United States any more. They often have better opportunities in their home countries than they have in the U.S. We can't take it for granted that everyone wants to come here, we have to start competing for the best global talent.

Entrepreneurship in India and China: Catalyzed by Returnees

In China, I also met with local entrepreneurs in an effort to get an update on the local entrepreneurship scene. I have been travelling frequently to China and India over the past few years to research their education programs for engineers as well as their entrepreneurship ecosystems. I also observed the impact that returnees from the U.S. are having on the local economies. In a nutshell, I learned that entrepreneurship is exploding in both countries and that they are beginning to innovate like we do.

In technology entrepreneurship, success comes after several attempts at starting companies. In both China and India, there used to be such a strong taboo associated with failure and such low social esteem granted to start-ups that parents would discourage their children from becoming entrepreneurs. Failed entrepreneurs were considered outcasts and would not be given a second chance.

This is rapidly changing. Chinese and Indian youth now have role models as a result of success of the first generation of technology start-ups, a success that has encouraged acceptance by their parents of entrepreneurial risk. Prospective entrepreneurs are also much more ambitious and confident than their parents were, and they connect with each other and their counterparts in other parts of the world through social networks.

The most important catalyst of entrepreneurship in China and India is the tide of returnees from the West—particularly from the U.S. Tens of thousands of highly educated and experienced entrepreneurs, along with students from top universities, have been returning home over the past few years and teaching locals about the ways of the West. They have been causing a rapid change in local cultural values and fertilizing the entrepreneurship landscape. They are building bridges to the West via social networks.

If you visit the start-up incubators in Beijing or Bangalore or attend technology start-up events, you find that 30 percent to 40 percent of the start-ups have returnee founders. These returnees are teaching locals how to build world-class companies and how to innovate. In almost every high-growth tech company in China, you find returnees in senior management positions. In scientific research, top research labs have returnees in lead positions. And these scientists are beginning to make breakthroughs. They are acting as a catalyst for innovation and economic growth in China and India.

This is a good thing for India and China and will produce long-term dividends for America by creating a two-way "brain circulation." It will expand American markets and spread American values—both are also good things. But the greatest economic growth will be in India and China. There is a high likelihood that Google-class companies will emerge from those countries instead of from the U.S. and that Silicon Valley will, for the first time, face unprecedented competition.

Many people believe that America is the most entrepreneurial land in the world, that it provides better opportunities for entrepreneurs than can countries such as India and China. To learn more about the entrepreneurship scene and how returnees from the U.S. are faring once back home in China and India, my team at Duke, UC-Berkeley, and Harvard surveyed 153 skilled immigrants who had returned to India to start companies and 111 who had returned to China. The title of the paper (again published by Kauffman Foundation) tells the story: *The Grass is Indeed Greener in India and China for Returnee Entrepreneurs: America's New Immigrant Entrepreneurs, Part 6*. Here is what we learned:

- The most significant factors drawing both Indians and Chinese entrepreneurs home were economic opportunities, access to local markets, and family ties. More than 60 percent of Indian and 90 percent of Chinese entrepreneurs said that the availability of economic opportunities in their countries had been a major factor in their return. Seventy-eight percent of Chinese entrepreneurs, and 53 percent of

Indian ones, had been lured by the attraction of local markets. And 76 percent of Indian entrepreneurs and 51 percent of Chinese entrepreneurs cited family ties as a factor that had brought them back home.

- Surprisingly, 72 percent of Indian and 81 percent of Chinese returnees said that the opportunities to start their own businesses were better in their home countries. The majority of entrepreneurs (54 percent of Indian, 68 percent of Chinese) found professional growth faster there than in the U.S. And for most (for 56 percent of Indian and 59 percent of Chinese) returnees, the quality of life was better than — or at least equal to — what they'd enjoyed in the United States.

What does this mean? It means the U.S. doesn't have the advantage in entrepreneurship that some people believe it does. We are going to have to compete to attract the world's best entrepreneurs—people such as the legendary venture capitalist Vinod Khosla and Google founder Sergey Brin.

The First American Brain Drain and Effect of Visa and Per-Country Limits

The U.S. has always been a land of immigrants. It has historically benefitted from an outflow of talent from the rest of the world. America has never experienced a brain drain and does not even recognize its symptoms. But, just as there are millions of people around the world trying to come to the United States, there is an outflow of highly skilled talent in progress that is fuelling the economic growth of countries such as India and China.

In 2006, my research team at Duke and UC-Berkeley conducted a survey of 2,053 technology and engineering firms founded nationwide in the period from 1995 to 2005. We found that 25.3 percent had a chief executive or lead technologist who was foreign born. We estimated that in 2005, immigrant-founded tech companies generated $52 billion in revenue and employed 450,000 workers.

We learned that the majority of immigrant entrepreneurs—who start 52 percent of Silicon Valley's companies—came to the U.S. to study. On average, they started their companies 13 years after their arrival in the U.S. We determined that, in 2006, foreign nationals residing in the U.S. were named as inventors or co-inventors in 25.6 percent of World Intellectual property Organization (WIPO) patent applications filed from the U.S., and immigrants had been critical to the success of some of America's largest companies. For example, they contributed to 72 percent of the total patent filings at Qualcomm, 65 percent of the total at Merck, 64 percent of the total at General Electric, and 60 percent at Cisco Systems. More than 40 percent of the international patent applications filed by the U.S. Government also had foreign national authors.

We were puzzled as to why foreign-national patent filings had increased so dramatically—by 337 percent in eight years. To explain this increase and understand the correlation with immigration trends, we developed a methodology to estimate the inventors' countries of origin. No such data are available from the U.S. State Department or the Citizenship and Immigration Services (USCIS).

What we found was shocking. As of September 30, 2006, there were 500,040 principals in the main employment-based categories and an additional 555,044 family members awaiting legal permanent-resident status in the United States. The backlog had been building since the mid'90s.

The reason for the increasing backlog is that only 120,000 visas are available per year in the key visa categories for skilled workers, with no more than 7 percent of them to be allocated to immigrants from any one country. So, immigrants from populous countries such as India and China have the same number of visas (8,400) available as those from low-population countries such as Iceland and or Costa Rica.

The "New Immigrant Survey" — a nationally representative longitudinal study of new legal immigrants — collected extensive data on the immigrant cohort of 2003. It found that the process of applying for permanent residence is so arduous that approximately 17.4 percent of new legal immigrants became depressed as a result of the visa process. Approximately 21.7 percent of new legal immigrants and 34.5 percent of "employment principals" either plan to leave the United States or are uncertain about remaining.

Based on the long and growing queue and the percentage of immigrants who felt aggrieved by the immigration process, we concluded that the potential exists for a sizeable reverse migration of skilled workers from the U.S. to their home countries or other countries, such as Canada, that welcome them. In August 2007, Kauffman Foundation published our paper titled "Intellectual Property, the Immigration Backlog, and a Reverse Brain-Drain: America's New Immigrant Entrepreneurs, Part III," which details these issues and our predictions.

Indeed, dozens of front page articles in major newspapers, CBS and NBC TV prime-time segments, reports by the Chinese government, and our visits to India and China have now substantiated our fears that students and skilled workers are returning home in record numbers, and the trend is accelerating.

A new research report by Stuart Anderson Executive Director of the National Foundation for American Policy, determines that the backlog is particularly severe for Indian nationals. According to this report, a highly skilled Indian national sponsored today for an employment-based immigrant visa in the third preference could wait for 70 years to receive a green card (this number is based on dividing an estimate of 210,000 or more Indians waiting for EB-3 visas by 2,800—the number of Indian professionals who receive permanent residence in this category each year).

What does this mean? This means the U.S. is giving an unintentional gift to China and India by causing highly educated and skilled workers, frustrated by long waits for visas, to return home. We are exporting our growth and competitiveness.

How Can We Reverse the Tide?

There is no way to put the genie back in the bottle, but we can give Silicon Valley and America's hi-tech industries a fighting chance to compete globally by enabling them to retain the skilled immigrants that are working for them, and to hire the people that want to work for them. Students may say they want to return home, but once they have worked in American industry and founded their start-ups, it becomes increasingly difficult to do so. The million skilled workers and their families who are waiting for green cards have already made a decision that they want to stay in the U.S. permanently. The only thing holding them back is the U.S. government. Let's not force these doctors, scientists, engineers, and researchers to leave by delaying their visas.

1. The right solution is to significantly increase the numbers of visas that are offered to skilled workers in the EB1-through-3 categories. We also need to remove the per-country limits. If a move such as this proves politically untenable, then the conversion of temporary visas to permanent residencies could be tied to the purchase of a house, of say $250,000 or more in value.

2. We should provide permanent-resident visas for graduates of top U.S. colleges. Given that, among U.S. postgraduate engineering and science students, nearly half of masters and most PhD students are foreign nationals, it makes sense to encourage these students to stay in the U.S. after graduation. Though it will not guarantee that they will stay, it will certainly make it more likely.

   To limit abuse of this program, it should only apply to degree holders from research universities or universities with established and well-regarded science, technology, engineering and mathematics (STEM) programs. We should also require that students receive job offers from legitimate U.S. corporations.
3. Another solution is to allow skilled workers to get a green card if they start a company that employs Americans. As we noted from our research, the majority of the foreign-born entrepreneurs who started Silicon Valley companies entered the U.S. for education or work. They started companies 13 years, on average, after their arrival in the U.S. So, this was the cohort that had entered the U.S. in the '80s and early '90s. A sizeable proportion of the 2000 cohort is stuck in "immigration limbo." There are tens of thousands of such immigrants who are ready to start companies that create jobs if we let them.

Why does this matter? It matters because we are now in a knowledge economy—in which in skilled talent plays a vital role. We face brutal competition from all over the world. Other countries have learned to play our game and they have the advantage of larger populations. We want their best and brightest scientists and engineers playing on our team and making us more competitive.

In: STEM Workforce Trends ... ISBN: 978-1-62257-818-4
Editors: C. Martin Garcia and S. Cook

Chapter 8

STATEMENT OF LINDSAY LOWELL, DIRECTOR OF POLICY STUDIES, INSTITUTE FOR THE STUDY OF INTERNATIONAL MIGRATION. HEARING ON "STEM THE TIDE: SHOULD AMERICA TRY TO PREVENT AN EXODUS OF FOREIGN GRADUATES OF U.S. UNIVERSITIES WITH ADVANCED SCIENCE DEGREES?"*

I would like to thank Chairman Gallegly and the Members of the Subcommittee for inviting me to testify here today. I have studied the global mobility of highly skilled workers for two and a half decades from the vantage of policy, demography, and labor markets. Policy changes are needed to improve America's competitive advantages, but perhaps in different directions than those under current debate.

There is little doubt that immigrants benefit the U.S. economy, especially those in the S&E workforce which is the backbone of our information age economy. I am a third generation American and was raised with my Romanian grandmother, a nurse, who pushed her son to excel. My father was a Biochemist who started a private laboratory to develop new technologies to test for disease and he held several patents. His entrepreneurial drive makes it easy for me to visualize how the immigrant experience benefits us all and it

* This is an edited, reformatted and augmented version of the testimony given on October 5, 2011 before the House Committee on the Judiciary, Subcommittee on Immigration.

makes me aware of the uniqueness of such individuals. I was also raised in southern California and during NASA's race to space and the effects of later downscaling of the aerospace industry. There is little disagreement about on the value of attracting S&E immigrants, but we should make a distinction between the supply and demand-side.

I believe that demand side policies will, in the long run, be more successful in building the American economy, and benefitting domestic workers, than finding ways to expand the immigration of S&E students and workers. I come to this conclusion, not only because it accords with my experience and logic, but because data supports that belief. Despite commonly held views about the declining abilities of ever fewer S&E students, I think the data shows that the S&E pipeline is far from broken. Despite the common assertion that there is a shortage of S&E workers, I think the data indicates a loose labor market with a ready labor supply. And despite the belief that more is better, or that streamlining visa processing is the best response to our cumbersome system, I think the challenge to constructing a competitive immigration policy is creating incentive-driven regulations to admit the best and brightest.

THE DOMESTIC STUDENT PIPELINE ISN'T BROKEN

The well-known "Rising Storm" reports from the National Academies investigate America's future competitiveness, raising a number of concerns, and offering several recommendations. They and others question the strength of the domestic S&E pipeline. By domestic I mean the native born and the already resident foreign born, both adults and those who arrive as children. By pipeline, I refer to the long pathway up through high school, to college, and ultimately along the S&E career path.

My colleague Hal Salzman and I decided to look into the strength of the S&E pipeline. We were somewhat surprised to find many concerns misplaced or wrong. We explore broad "core" S&E occupations—the natural sciences, engineering and information technology—because they share aptitudes and skills. Here the issue is whether there is a problem that calls for demand side and government interventions. While there are specific fields in which we observe hiring (demand) outpacing supply, this tends to be short-lived as supply is surprisingly responsive: take the case of petro engineers where the number of *domestic* graduates more than doubled in a few years in response to increased salaries. Our research finds that the S&E pipeline is reasonably strong even if it can and likely should be improved. Professor Salzman presented our findings to the House Subcommittee on Technology and Innovation in 2007 and I update them here.

Large Student Body, Small S&E Workforce

There pool of students with the ability to purse an S&E career is far larger than the S&E workforce. The S&E workforce is fairly small at roughly 5 percent of the U.S. labor force. The *number* of students who score high on math or science is large, and we graduate more S&E students that annual S&E workforce growth. The Bureau of Labor Statistics projects 190,000 annual S&E job openings due to growth *and* replacement needs between 2008 and 2018. Annually, between 1995 and 2007 there was an increasing number of *domestic* S&E graduates, averaging 408,000 bachelors; 78,000 master; and 21,000 doctoral graduates for a total of 507,000. The rate of growth of domestic graduates will lessen as our population ages, but there will be no sharp decline.

International Tests Averages Can Mislead about U.S. Performance

It is true that U.S. students regularly place mid-way in the international pack on math, albeit they perform better on science and reading. Yet, the age groups sampled internationally are not fully comparable and the ranking between countries is often not statistically significant. And as a separate sample in Massachusetts of the Trends in International Mathematics and Science Study (TIMSS) recently demonstrated, students in *individual* U.S. states test better than those in many of the smaller nations that rank above the *average* American. Furthermore, we are a large nation and we have a lot of students in the upper tail of test score performance.

National Tests Show Improved Math Scores

The SAT math scores of college bound high school seniors have increased steadily since 1980 after a marked dip in the 1970s. To be sure SAT math scores have recently stabilized at a level a little lower than their 2005 peak, but that is not yet a trend and may simply reflect an increase in the proportion of high school test takers. What is a trend are the smart improvements over the past two decades in the NAEP math scores of 8th graders and the steady trend upward of 12th grade NAEP math scores for all groups.

Student Interest in S&E Has Been Steady for Four Decades

Surveys of incoming college freshman by the Higher Education Research Institute (HERI) since the early 1970s have found about one third express an interest in pursuing a S&E field of study. Additionally, Professor Salzman and I analyzed three decades of longitudinal data and found pretty much steady rates through the pipeline. That is to say there has been little change in the percent of entering high school students who graduate with an S&E bachelor or advanced degree; or the percent who go on to an S&E job. We do not find evidence that the student pipeline is at imminent risk of failure, that it is grossly deficient across the board compared with other nations, or most importantly that it produces too few domestic students for the S&E workforce. Student interest in science and engineering has remained strong and leads to a substantial supply of S&E graduates.

The S&E Labor Market Is Not "Tight"

Most Economists will not readily admit that anything like a shortage of labor exists. But that is not quite right; theory says that given strong demand an existing shortage will right itself with time. Richard Freeman has shown that engineering markets have oscillated as increasing wages induce an increased supply of students and, in turn, rebounding supply lowers wages and student enrollments again decline. That is, there are short term or cyclic shortages and longer-term, structural shortages. Cyclic shortages are what temporary visas might fill until the market rights itself, while permanent migration offsets longer run structural shortages. More to the point, wages settle to a level that reflects the relative balance of supply and demand. If demand outstrips supply, wages escalate to address the shortage of labor. If there is a plentiful supply of labor, wages will be lowered. There is evidence that there is a substantial supply in the S&E labor market and that it is best characterized as loose (not tight).

Poor Retention of S&E Workers

Historically, the roughly 5 million S&E workforce is one third of the roughly 16 million workers who hold at least one S&E degree. There are far more S&E trained workers than the number employed in S&E occupations. Of course, S&E graduates may find related jobs outside of narrowly defined S&E occupations. My colleagues and I have explored a broader definition of "S&E jobs." We start with core S&E occupations but also include as related those "non-S&E" occupations in which S&E trained workers find themselves namely medical jobs, S&E management, and jobs workers report as closely related to their education. Even with this broadened definition, about one-fifth of early career S&E graduates are in jobs that do not use their training and that increases to one third by mid-career. It is important to ask why retention is poor.

The Boomlet of S&E Immigrants

The Immigration Act of 1990 boosted the admission of highly skilled migrants. That shows up clearly as the foreign born today are roughly half of PhDs and about one fifth of master degree holders less than 45 years of age, but less than one-seventh of all older workers. We need not review the boom of immigrants in S&E fields of study, or as post-docs, or in S&E occupations. The high percent of immigrants, however, does not *ipso facto* signal a shortage of natives as some assert, the high percent may have a number of causes. Consider the independent demand for students by colleges or the dynamics of a globalized market for students and workers. In the worst scenario, immigrants may discourage the pursuit of S&E careers, for which there is limited evidence but evidence just the same. What we can clearly say is that the immigration boomlet attests that policy is not *numerically* "restrictive" and that it reinforces an impression that immigration is a significant addition to available labor

S&E Wages Lag "Alternative" Professional Jobs

Some observers argue that S&E pays less, or that the wages increase slower, than "other" jobs. If S&E pays less, it suggests that other highly-skilled professionals are in greater demand and/or other factors such as a ready supply of labor are depressing S&E wages. Yet; the Department of Commerce reports that S&E occupations pay better than *all other* jobs, but that is not a comparison with jobs that compete with the S&E pipeline. Even so, *on average* S&E jobs pay more than the *average* paid to all other *professionals* in non-S&E jobs. What then is the right type of wage comparison?

Like a good researcher, I believe we need more studies; here is some of what we do know. Thomas Espenshade finds that the earnings of S&E occupations increased rapidly from 1970 to 1990 only to see wage growth fall sharply in the 1990s, which he speculates may be related to increasing immigration. Thomas Lemieux finds that core STEM occupations experienced slower wage growth than other professional jobs through the early 2000s, especially those of lawyers and medical practitioners, which he speculates may be due to outsourcing. Clair Brown's research echoes some of these speculations for computer engineering although she is cautious about predicting the future.

What specific "alternative" professional careers appeal to individuals in the S&E pipeline? In the HERI surveys, focusing on college freshman who intend to pursue S&E studies, we find these freshmen report their preferred non-S&E careers to be medical practitioner, veterinarian, teacher (secondary), business executive-manager-administrator, lab technician or hygienist, lawyer or judge, therapist, accountant, and pharmacist or architect. Using U.S. Census data I find that core S&E occupations paid better and had faster wage growth than these non-S&E alternatives after WWII through the Sputnik and cold war decades, but they increasingly paid less than these alternative jobs by 2000 and 2009. This is consistent with the impression that there is not a labor shortage in today's S&E market, but rather it is loose.

The Best and Brightest Are Not about More and It's Not Easy

Of course, job supply is not a zero sum game and the supply of immigrants might boost innovation, productivity and the job prospects of domestic workers. Casually read most articles on this subject and they tell us that immigrants are more likely to be Nobel Prize winners and they start 25 percent or more of businesses in Silicon Valley or other high-tech ventures. William Kerr finds that H-1Bs boost patenting by Indian and Chinese inventors. Jennifer Hunt finds that the foreign born outperform natives on academic publishing, patents and their commercialization, wages, and in starting businesses. Others find that highly skilled immigrants are associated with favorable effects on economic growth.

These facts are sound, but they are not the entire story. We need to go beyond a handful of studies and, with additional data and research, ask what increasing the number of immigrants might achieve. In particular, let us consider the oft-cited study of immigrants and patents/citations which, as this committee knows, are a critical but not final harbinger of innovation in the marketplace. Moreover, empirical results at the margin do not readily extrapolate beyond the mean. In other words, if the measured impact of a 1 percent increase in skilled immigration increases patenting by 5 percent; it does not follow that a tenfold increase in immigration will generate 50 times the number of patents.

Most importantly, though less well understood, is the nature of "selectivity" in immigration patterns and flows. International mobility is always a story of selectivity, or the number of migrants who are actually the best and the brightest. Not all are. The OECD finds a mildly adverse impact of large numbers of immigrants on migrant selectivity. What is more, globalization by lowering the bar to mobility, and by creating new opportunities abroad, should lead to lower selectivity over time, i.e., a smaller and smaller proportion of migrants with that certain X-factor that generates innovation or entrepreneurial activity. In a competitive, globalizing world getting the immigrants with that X-factor should be what innovative admission policy is all about; not fighting yesterday's battles over more or less immigration.

The Effects of Globalization on Nobel Prize Winners

Hunter, Oswald and Charlton examine the mobility of world-class scientists. They find, encouragingly, that the U.S. is one of the world's top destinations. Yet, these migrants are no more productive than domestic scientists after they migrate. And over the past half century they have been a decreasing percentage of America's Nobel Prize winners. What is more, as the immigrant proportion of Nobel Prize winners has decreased, the proportion of immigrants with an S&E PhD has increased. Simply admitting more PhDs has no effect on getting this very special X-factor. Then Nobel Prize winners are, we all should acknowledge, a very tiny number. The point, nevertheless, is that globalization works against selectivity.

It Takes Time to Grow an Entrepreneur

Immigrant entrepreneurs are a long term bet. Vivek Wadhwa reports from his specialized, high tech sample that immigrant entrepreneurs had been in the country at least 13 years. David Hart reports from his sample of fast-growth firms that immigrant entrepreneurs had been in the country 26 years on average. In other words, the selectivity process for entrepreneurs operates through drive and experience. What we know of the process suggests they came by their drive after establishing themselves and not before. Some were, indeed, previously students before becoming greencard holders. But selectivity operates here as well. Consider the difficulties that Australia has had with awarding permanent residency to foreign graduates: incentives are muddied when an award of permanency is added to the core purpose of getting a degree. The United States retains two-thirds of foreign PhD graduates and we have no solid data at all on whether or not rates of retention are changing. Students who are primarily motivated by their field of study are those most likely to excel and become entrepreneurs.

Innovation and Entrepreneurship

Immigrants are rightly lauded for starting high tech companies. Individual examples abound, but as impressive as they are one should acknowledge they are individual anecdotes. The problem with the "immigrant as entrepreneur" story is that it relies on both anecdote and surveys provide a surprisingly limited basis for generalization. Impressively, Anna Lee Saxenian famously reported that Chinese and Indian computer scientists and engineers ran one-quarter of Silicon Valley's high-tech firms in 1998. Vivek Wadhwa found that at least one immigrant was a founder of one-quarter of all engineering and technology companies between 1995 and 2005 and up to half in Silicon Valley. Stuart Anderson finds that immigrants started one-quarter of venture-backed companies over the past 15 years. Yet, Hart et al. find that immigrants nationally were part of the founding team of a far smaller one-sixth of high-growth companies. Academics worry over what types of companies are included in these special-purpose samples and how comparable and representative they are.

Consider that Census data shows that no better than 5 percent of all workers in S&E occupations are self-employed, perhaps unsurprisingly, less than the percent of workers in other professional jobs. The S&E foreign-born, in turn, are less likely than S&E natives to be self-employed. On the other hand, in high technology industries nationally a little over one-sixth of the self-employed are immigrants, closer to Hart et al.'s figures. Again more impressively, close to three-tenths of the self-employed in IT industries are immigrants in Silicon Valley. Impressive, yes, but the valley's share of immigrants in the S&E workforce is roughly twice as large. In other words, one has to admit a rather large number of immigrants to generate what is evidently a very concentrated pattern of immigrant entrepreneurship in key metropolitan areas.

Furthermore, it is not clear that it is immigration that provides the most critical ingredients for innovation "hot spots." In *The Silicon Valley Edge*, Lee and his colleagues argue that there are 10 reinforcing factors behind the valley's success, of which immigrants are only one part. Success in innovation, in the final analysis, is about creating the right business, intellectual, regulatory and academic environments; not about supply-side economics.

SUMMARY

A generous number of S&E migrants has been admitted to the U.S. and, as the system/policy is currently structured, will continue to admit large numbers of immigrants. At the same time, there is little evidence that our educational pipeline produces too few domestic students able and willing to pursue an S&E career. And employment opportunities in most S&E occupations, particularly due to two recent recessions, are not as strong as they should be to attract more domestic workers. Economists tell us that tighter labor markets induce wage gains and, in turn, more domestic residents would be retained all through the pipeline.

Today's admission system is faulty, however, and shy of a complete redesign, targeted changes should be made. I suggest three principles that should inform policy and regulatory changes.

First, changes should be careful not to artificially induce significantly increased admissions of foreign S&E workers. America's competitive advantage is best served by spurring domestic demand. It is not a good idea to create new visas or expanded caps, escalating caps, or cap exemptions. Nor is it a good idea to award automatic greencards which has the additional downside of creating the wrong incentives to, for example, pursue specialized education in the United States. Neither does it make sense to expand temporary programs, particularly those with long stays and no screening for intent to stay, without a corresponding capacity to permit adjustments to permanent status.

Second, uniquely innovative people are not common and policy should be selective; it should use incentives to admit immigrants who are the best and the brightest. Getting this formulation right is a difficult quest as we know so little about what works. Comparisons to systems in other countries are problematic not only because other countries are not first in skilled migrants preference queue as we are, but also because evidence does not support casual assumptions that point systems say work better than our system. A different system would keep employers in the driver's seat but provide different mechanisms for admitting migrants. For example, the Jordan Commission and, more recently, economists have weighed in favor of market-oriented admissions. One idea is to try truly temporary 2-3 year visas that are awarded in quarterly and spot auction markets on a pilot basis. Entrepreneur visas, that require benchmark investments for new graduates or temporary visaholders, are another good idea. And, given the evidence, we might want to target those with a track record than casting a wide, loose net for young graduates who might, someday, become successful entrepreneurs.

Third, policy should be as fair as possible. Today's system has put many good candidates in an admission backlog, while frustrating employers' need for timely resolution of workers' employment status. There is a sizeable backlog of greencard applicants for employment based visas. We and their employers have invested in these workers' skills and they desire to remain and work in the United States. Clearing that backlog, allocating unused employment visas from past years or by some other means, makes sense. There are many ideas on how to improve policy, in which the Committee is well versed, and my purpose here is not to detail specific recommendations but rather to present evidence for making informed decisions and a few principles to frame those decisions.

INDEX

D

E

F

G

H

I

J

K

L

M

N

O

P

Q

R

S

T

U

V

W

X

Y

Z